BBC

GCSE BITESIZE revision

ography

e Freeman
Rae
Balderstone

BC Worldwide Ltd,
Wood Lane, London W12 0TT

2002; reprinted 2003, March 2004

an, Nicola Rae, David Balderstone/BBC Worldwide Ltd, 2002
ed

26 X

ction by Tien Wah Press Pte Ltd, Singapore

nd by Tien Wah Press Pte Ltd, Singapore

Page make-up by Oxford Designers & Illustrators
Illustrations © Peter Bull, 2002

Contents

About Bitesize

GCSE Bitesize is a revision service designed to help you achieve success at GCSE. There are books, television programmes and a website, each of which provides a separate resource designed to help you get the best results.

TV programmes are available on video through your school, or you can find out transmission times by calling 08700 100 222.

The website can be found at
http://www.bbc.co.uk/schools/gcsebitesize

About this book

This book is your all-in-one revision companion for GCSE. It gives you the three things you need for succesful revision:

1 **Every topic is clearly organised and clearly explained.**

2 **The most important facts and ideas are highlighted for quick checking:** in each topic and in the extra sections at the end of the book.

3 **There is all the practice you need:** in the 'check' questions in the margins, in the practice sections at the end of each topic, and in the exam questions section at the end of the book.

Each topic is organised in the same way:

■ **The bare bones** - a summary of the main points, an introduction to the topic, and a good way to check what you know.

■ **Key facts** highlighted throughout.

■ **Check questions** in the margin - have you understood this bit?

■ **Remember tips** in the margin - extra advice on this section of the topic.

■ **Exam tips in red** - specific things to bear in mind for the exam.

■ **Practice questions** at the end of each topic - a range of questions to check your understanding.

The extra sections at the back of this book will help you to check your progress and to be confident that you know your stuff:

Exam questions and model answers

■ A selection of exam questions with the model answers explained to help you get full marks.

About this book *continued*

Topic checker

- Quick questions in all topic areas.

- As you revise a set of topics, see if you can answer these questions - put ticks or crosses next to them.

- The next time you revise those topics, try the questions again.

- Do this until you've got a column of ticks.

Last-minute learner

- The most important facts in just six pages.

Answers at the end of the book.

About GCSE Geography exams

- **Knowledge** This means learning some facts about places, geographical ideas, processes and issues that you have studied. You'll need to be able to name particular examples and say where they are. Make sure you know what geographical terms such as **erosion** and **deposition** mean, too.

- **Understanding** The examiners want you to show that you know how certain geographical processes work and can explain different geographical features and patterns.

- **Geographical skills** That's using maps, graphs, photographs, statistics, and so on. You might need to draw a diagram or sketch map to help you explain an answer. You will be provided with various resources and the GCSE questions will try to find out whether you can analyse and interpret them. Don't forget the fieldwork skills you have used to collect, present and analyse information.

- **Attitudes and values** Different groups of people have different views about geographical issues. Some GCSE questions will try to find out whether you understand these views and why different groups of people hold them.

Revision tip – making revision notes

One of the main functions of revision in geography is to help you learn important information about geographical ideas, processes, case studies and examples that you have studied. Making revision notes will help you to rearrange your class work and other work so there's a manageable amount of information to learn, organised in a way that makes it easier to learn. Below is an example to show you how revision notes could be organised:

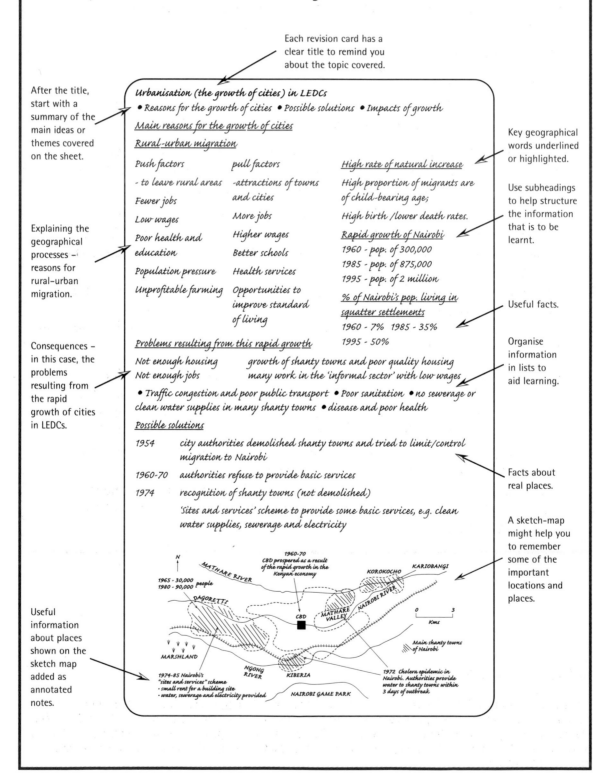

Each revision card has a clear title to remind you about the topic covered.

After the title, start with a summary of the main ideas or themes covered on the sheet.

Explaining the geographical processes – reasons for rural–urban migration.

Consequences – in this case, the problems resulting from the rapid growth of cities in LEDCs.

Useful information about places shown on the sketch map added as annotated notes.

Key geographical words underlined or highlighted.

Use subheadings to help structure the information that is to be learnt.

Useful facts.

Organise information in lists to aid learning.

Facts about real places.

A sketch-map might help you to remember some of the important locations and places.

Urbanisation (the growth of cities) in LEDCs
- Reasons for the growth of cities • Possible solutions • Impacts of growth

Main reasons for the growth of cities

Rural-urban migration

Push factors	pull factors	*High rate of natural increase*
- to leave rural areas	-attractions of towns and cities	High proportion of migrants are of child-bearing age;
Fewer jobs		High birth /lower death rates.
Low wages	More jobs	*Rapid growth of Nairobi*
Poor health and education	Higher wages	1960 - pop. of 300,000
	Better schools	1985 - pop. of 875,000
Population pressure	Health services	1995 - pop. of 2 million
Unprofitable farming	Opportunities to improve standard of living	*% of Nairobi's pop. living in squatter settlements*
		1960 - 7% 1985 - 35%
		1995 - 50%

Problems resulting from this rapid growth

Not enough housing   growth of shanty towns and poor quality housing
Not enough jobs   many work in the 'informal sector' with low wages

• Traffic congestion and poor public transport • Poor sanitation • no sewerage or clean water supplies in many shanty towns • disease and poor health

Possible solutions

1954   city authorities demolished shanty towns and tried to limit/control migration to Nairobi

1960-70   authorities refuse to provide basic services

1974   recognition of shanty towns (not demolished)

        'Sites and services' scheme to provide some basic services, e.g. clean water supplies, sewerage and electricity

1965 - 30,000
1980 - 90,000 people

1960-70
CBD prospered as a result of the rapid growth in the Kenyan economy

MATHARE RIVER

KOROKOCHO   KARIOBANGI

DAGORETTI

MATHARE VALLEY   NAIROBI RIVER

CBD

0   3
Kms

MARSHLAND

Main shanty towns of Nairobi

NGONG RIVER   KIBERIA

1974-85 Nairobi's "sites and services" scheme
- small rent for a building site
- water, sewerage and electricity provided

NAIROBI GAME PARK

1972 Cholera epidemic in Nairobi. Authorities provide water to shanty towns within 3 days of outbreak

Essential terms and concepts in GCSE Geography

The terms and ideas explained below are referred to throughout this book. They are very important in GCSE Geography and you need to make sure you learn them as part of your revision. Understanding and using the correct geographical terms in your exam can ensure that you gain valuable marks.

Urban - A built-up area, e.g. a town or city.

Rural - An area of open land and generally low-population density, i.e. the countryside.

LEDC - A less Economically Developed Country.

MEDC - A more Economically Developed Country.

Geographical patterns - Trends or patterns found across different parts of the world. Geographical patterns tell us what is happening in the world.

Geographical processes - Geographical processes help to explain the world in which we live. For example, geographical processes can be used to explain how different landscapes are formed or why a city has grown in a particular way.

Globalisation - This is the process whereby people and places are becoming increasingly interconnected. As technology develops, the world seems to be getting smaller. People can now travel great distances at high speeds and the Internet allows us to make instant contact with distant people and places.
The process of globalisation has had an important effect on trade and production. Many individual countries and nations have become part of a single global economy. A financial or commercial setback in one country can have an impact on the whole world trade system.

Multi-National Company (MNC) - As part of the process of globalisation, a growing number of large companies now operate in many different countries. These companies operate beyond national boundaries and are therefore multi-national. Generally, the headquarters of MNCs tend to be located in MEDCs, while the production centres (manufacturing centres) tend to be located in LEDCs.

Sustainable development - This is an important concept in geography. If something is sustainable it can be kept going. Sustainable development meets people's present needs while conserving the natural environment for future generations.

Acknowledgements

Alamy, p. 102; Corbis/Owen Franken, p. 14; Corbis/Tiziana and Gianni Baldizzone, p. 57; Getty Images, pp. 65 and 95; Holt Studios, pp. 94 (top and bottom); London Aerial Photo Library, pp. 47 and 91; Lonely Planet Images, pp. 109 (bottom) and 116; Lonely Planet/Dave Lewis, p. 116; Lonely Planet/Richard I'Anson, p. 45; MEG/D. Brunner, p. 141; Merry Hill Centre, p. 133; Meteosat. pp. 15 and 61; Ordnance Survey, pp. 11, 136 and 139.

THE BARE BONES
➤ There are many different types of maps in geography.
➤ In your exam you will need to be able to read and interpret maps.

A Types of map

1 Here are some of the different types of map that may appear in your exam:

> **Ordnance Survey maps (OS maps)** are detailed maps produced for the UK. They come in a **standard format**.

> **Sketch-maps** are **simple** maps that are drawn quickly and **not** usually to **scale**.

> **Statistical maps** are maps that show **statistical information** (e.g. birth rates in different countries of the world). A common statistical map is a **choropleth** map (see p.72). A **choropleth** map uses **shading** to show different **values** or groups of values.

2 The rest of this section focuses on Ordnance Survey (OS) maps. However, the skills that are explained here are relevant to nearly all maps.

Q Can you list three types of map?

B Grid references

1 Grid references are used to help people **locate** places or features on a map.

KEY FACT

2 Grid references can be given in two ways: <u>four-figure</u> grid references and <u>six-figure</u> grid references.

Remember
Grid references are a bit like co-ordinates, which you will have used in maths.

3 Four-figure grid references are used to locate a **single grid square**. For example, to locate 0722 on the grid opposite, you have to:

- Begin reading the map from the **bottom left-hand corner**.
- Read the numbers along the **bottom** of the map **first**.
- Move along the bottom row of numbers until you reach the number 07.
- The number 07 is found on the **left** side of the grid square.
- Now read the numbers on the **side** of the map.
- Move **up** the side of the map until you reach the number 22, which is at the bottom of the grid square.
- You will notice that an **L-shape** is formed in the **bottom left-hand corner** of grid square **0722**.

B

4 Six-figure grid references are used to locate the exact position of a place or feature **within** a grid square on a map. For example, to locate the village of Hampton on the grid below you must do the following:

- Imagine that each grid square has been **divided into tenths**.

- Again, read the map from the bottom left-hand corner.

- Move along the bottom row of numbers until you reach the left-hand side of the square with Hampton in it. The exact position of Hampton here is 06 and 3 tenths.

- Now move up the side of the grid until you reach the bottom of the grid square with the village of Hampton in it. The exact position of Hampton here is 22 and 7 tenths.

- The six-figure grid reference for Hampton is **063227**.

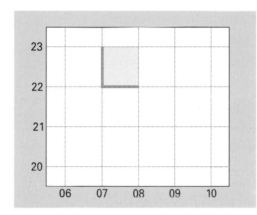

Four-figure grid reference.

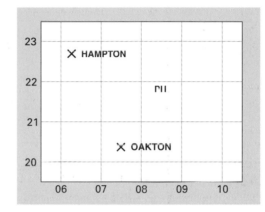

Six-figure grid reference.

Q Can you give the six-figure grid reference for the public house (on the grid opposite)?

C Scale

KEY FACT

1 The scale of a map helps you to <u>work out the distance</u> between one place and another.

2 On an **OS map**, scale is shown using a **scale line**, usually found at the bottom of the map or with the key.

3 In an exam you are likely to be given an OS map with a scale of **1:25000** or **1:50000**. On a 1:50 000 map, 1cm represents 50 000cm.

Q Do you know what 1cm represents on a 1:25 000 map?

PRACTICE

1 What is a choropleth map?

2 Using the map above, give the four-figure grid reference for the village of Oakton.

3 Using the map above, give the six-figure grid reference for the village of Oakton.

THE BARE BONES ➤ Understanding a map can involve using grid references, scale lines and map symbols and interpreting contour lines.

A Map symbols

KEY FACT ▷

1 To help <u>show detailed information</u>, most maps use symbols.

2 Map symbols are explained using a **key**. The key gives you the meaning of each symbol.

3 Different types of map may use different symbols. However, there are some common symbols that are used on most maps.

4 OS maps have a set of standard symbols. The OS map in this book has a key (p. 137). If you are given an OS map in an exam it will have a similar key attached. Make sure you use it to help you read and understand the map.

Q Can you draw and identify common OS map symbols?

B Relief

KEY FACT ▷

1 Relief is the shape of the land. Studying relief involves looking at the <u>height of the land</u>.

2 Relief is shown on a map using **contour lines** or **spot heights**.

3 Contour lines are brown lines that show the height of the land. The height of the land along a contour line is measured in **metres above sea level**.

• Contour lines join up points of **equal height** above sea level. The **height** of the land is the **same** at any **point along** one **single contour line**.

• Each contour line goes up in set intervals. On a 1:50 000 map each line goes up 10 metres.

• The closer together the contour lines, the steeper the relief of the land; the further apart the contour lines , the gentler the relief.

4 Spot heights are shown on a map as a dot, with the height of the land at that point written beside it.

Q What is the highest point on the contour map opposite?

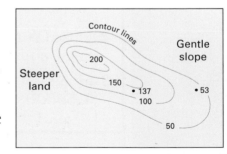

PRACTICE

Use the map of Newport opposite to answer the following questions:

1 Name the farm found in square 3585.

2 Name the features found at   a) 365829   b) 381834   c) 362878.

3 What is the height of the land at 388881?

B

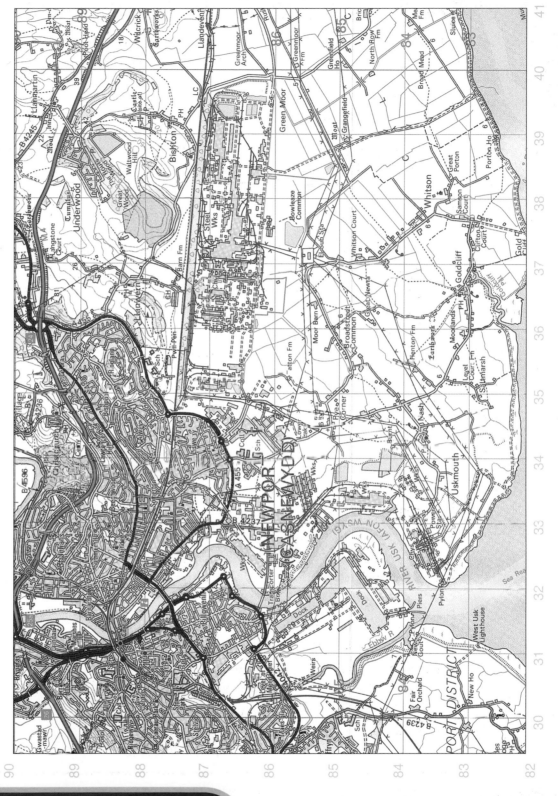

Ordnance Survey map of
Newport, South Wales

Drawing a simple sketch-map to illustrate a
case study is a useful exam technique.
Sketch-maps are quick to draw and can be
labelled to show information about a place.

THE BARE BONES

➤ Geographers often use data and statistics to help them discover patterns or trends.

➤ Geographical data is often shown in the form of a graph or chart.

➤ Geographers use many different types of graph or chart. Each one is designed to present a particular set of data in an appropriate way.

A Data and statistics in geography

KEY FACT

1 Geographers often use statistical data to <u>identify patterns or trends</u>.

2 Using statistics involves **analysing** numbers and figures. For example, population statistics include birth rate figures.

3 You are likely to be given relatively simple **statistical data** in your GCSE exam. You will be expected to **understand** and **interpret** the data you are given.

4 **Data** and **statistics** are usually shown in the form of a **graph**, **chart** or **table**. This section focuses on the different types of graphs and charts used in geography.

• In your exam you may be asked to **add** information to a graph.

Q Name three ways in which data and statistics can be presented.

B Line graphs

1 Line graphs are one of the simplest forms of graph used in geography. You will have come across these before in maths and science.

2 **Line graphs** tend to show **changes** over **time**. Time is **usually** shown on the x-axis.

3 If you are asked to **describe** the pattern on a graph, quote **figures** from the graph and give the relevant **year** or **month** relating to that figure.

4 The graph opposite shows changes in the population of **England and Wales** since 1700. It shows that the population has risen since the 1700s. In 1700 the population was 4.8 million.

It grew at a **steady rate** between 1700 and 1780, when it reached 6.4 million. From 1820 the population began to **rise rapidly**, reaching 59.4 million in 2000.

Remember
Learn the names of the two axes on a graph. The x-axis is horizontal. The y-axis is vertical: 'y to the sky!'

Q What was the population of England and Wales in 1990?

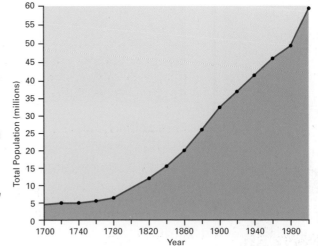

C Bar graphs and pie charts

Remember

When describing the pattern on a graph, quote figures and use descriptive words such as; rises, falls, declines, grows, rapidly, steadily.

Q What is the average rainfall for November in the Amazon rainforest?

1 **Bar graphs** use **shaded bars** of **different lengths** to show the value of a particular item of data.

2 There are a number of **different types** of bar graph used in geography:

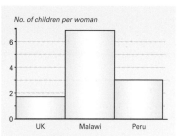

A bar graph is a basic way of displaying data. It shows instantly a pattern of data.

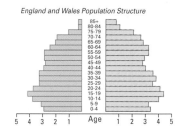

A population pyramid uses horizontal bars to show the age–sex structure of a population (see Population, p.75)

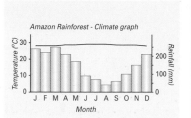

A climate graph is a bar graph and a line graph put together. Bars show monthly rainfall, the line graph shows temperature.

3 **Pie charts** and **divided bar charts** show **percentages** (see p.92).

D Scatter graphs

1 Scatter graphs are often used in geography to show the **relationship** between two sets of data.

2 Scatter graphs can be used to show how two sets of data **correlate** with each other. A **correlation** is a **link** between **two sets of data**. For example, you would expect to find a correlation between the birth rate of a country and the number of women using contraception.

3 You must not join up the points on a scatter graph. However, you can add a **line of best fit**.

Q Another graph that you may come across in geography is a triangular graph (see p.92). Can you find an example of one in your class notes?

A scatter graph showing a **strong positive** correlation between two sets of data.

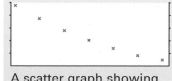

A scatter graph showing a **strong negative** correlation between two sets of data.

A scatter graph showing **no correlation** between two sets of data.

PRACTICE

1 What was the population of the UK in 1920?

2 Using the climate graph above, describe the climate of the Amazon rainforest. Discuss rainfall and temperature patterns in your answer, and quote figures from the graph.

Using images in geography

THE BARE BONES
- ➤ Photographs help people to gain a sense of what a place is like.
- ➤ Photographs can provide important evidence about geographical patterns and processes and show the impact of human activity on the environment.

A Photographs

1 Photographs are often used in GCSE Geography exams.

2 If a photograph is used in an exam question, it will have been included for a reason. Consider why the photograph has been used and try to <u>assess what information the examiner wants you to pick out</u> from the picture.

- The photograph below shows tourism development along the Mediterranean coast. It is surrounded by key information about such tourist resorts.

Remember
If you are asked to refer to a photograph in an exam question, make sure you use geographical terms and ideas in your answer.

The hotels are **clustered** together to form a resort.

High-rise hotel developments are very different in style to the traditional buildings found in the area. Many local governments in the region now have planning laws that regulate the height of hotels and other buildings.

Many hotels have **swimming pools**. They attract tourists to warm climate areas such as the Mediterranean, but they use a lot of water, which can drain local water supplies.

Around the tourist resort the **landscape** is relatively undeveloped and open.

Water sports are common in coastal resorts. They attract tourists, but they can also damage the environment through pollution. Vibrations and noise from the engines of boats may disturb animal wildlife.

Roads are needed to connect the tourist resort with other settlements and the airport.

3 Here are some of the questions you may be asked about the photograph above:

  a) Use **evidence** from the photograph to **describe** the **effects** of tourism on people and the environment.
  b) Use **evidence** from the photograph to give **two reasons** why people may be **attracted** to visit the Mediterranean coast.
  c) Use **evidence** from the photograph to **describe two ways** in which tourism may **change** the **landscape** of an area.
  d) Describe the **pattern** of tourism **development** in the **resort** shown above.

Q Can you list three typical features found in a coastal tourist resort?

A

4 You may be given an **aerial photograph** in your exam. Aerial photographs are featured on page 133 of this book.

5 You may be expected to use photographs in your GCSE Geography coursework. If you do include photographs in your coursework, make sure that you label them to show key information and refer to them in your writing; for example, 'the photograph above shows . . .'

B *Satellite images*

KEY FACT

1 Satellite images are often used in GCSE Geography exams.

2 Below is an example of a satellite image that you might see in an exam. The image was taken by the **Meteosat satellite**, which is positioned over the equator and **orbits** at the same **speed** as the **rotation** of the **earth**.

3 The Meteosat satellite gives a **broad picture** of the **earth**, showing **large cloud masses** in **low-pressure systems** and **clear skies** where there are **high-pressure systems**.

4 Not all satellite images used in exams will be related to the weather. In your exam you may be given a **satellite image** of a **place** or **region**, showing particular features such as **urban density**. These images **usually** have a **key** to help you **interpret** them, and may have grid lines marked on the image.

Try to give **co-ordinates** in your answer and use the **key** to help you **understand the image**.

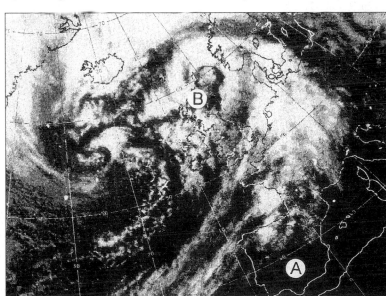

Meteosat image of Europe, 12 January 1994

Q Can you locate the UK on the Meteosat image opposite?

PRACTICE

1 a) List three land uses shown in the photograph of a Mediterranean tourist resort (see p.14).

b) List two possible sources of environmental damage or pollution.

2 Using the Meteosat image above, describe the pattern of cloud over Europe.

THE BARE BONES

➤ The earth is made up of four layers: the inner core, outer core, mantle and crust.

➤ The earth's crust is broken into sections called 'plates'.

➤ A plate boundary is the point where two plates meet.

➤ Most earthquakes and volcanoes occur close to or on plate boundaries.

A The structure of the earth

KEY FACT

Remember
The processes that take place within the earth often affect the processes that operate on the earth's surface.

1 The earth is made up of <u>four</u> layers.

• The **inner core** is the **centre** of the earth. It is made of **solid iron** and **nickel**. This is the hottest part of the earth, with temperatures of around **5500 °C**.

• The **outer core** is a part of the earth made up of **liquid iron** and **nickel**.

• **The mantle** is the **largest section** of the earth and is made up of **semi-molten rock** (half-melted). These **partially melted** rocks are called **magma**. Temperatures here are around **5000 °C**.

• **The crust** is the **thinnest** layer of the earth and is made up of **solid rock**. The crust is the outer layer of the earth, like the skin of an apple.

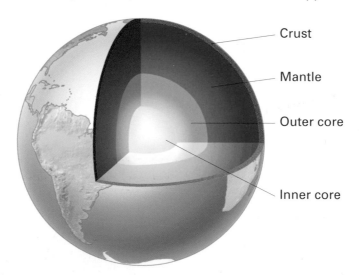

Crust

Mantle

Outer core

Inner core

2 The earth's crust is **broken up into pieces** called **plates**, which move or **'float' on the mantle**.

3 **Heat rises and falls** in the mantle creating **currents**. These are called **convection currents**.

• Convection currents cause the earth's plates to move. The plates move very slowly (around one or two millimetres a year).

• The movement of the earth's plates is known as **tectonic activity**.

Q Can you name the four main layers of the earth?

B The earth's plates

1 The movement of the earth's plates causes <u>earthquakes and volcanoes</u>.

2 The map below shows the world's main **tectonic plates**. The point at which two plates meet is called a **plate boundary**.

3 Most of the world's **earthquake and volcano zones are found on or near to plate boundaries**.

4 The earth's **plates move in different directions**. The direction of plate movement is shown on the map using arrows.

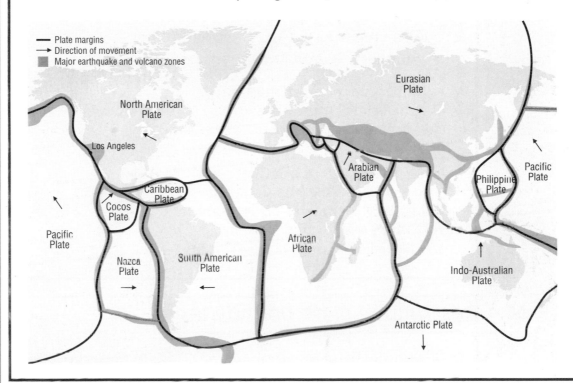

Legend:
— Plate margins
→ Direction of movement
■ Major earthquake and volcano zones

Plates shown: Eurasian Plate, North American Plate, Los Angeles, Arabian Plate, Pacific Plate, Philippine Plate, Caribbean Plate, Cocos Plate, Pacific Plate, African Plate, Nazca Plate, South American Plate, Indo-Australian Plate, Antarctic Plate

Q In what direction is the South American plate moving?

C Plate boundaries

There are <u>three</u> main types of plate boundary.

Q What happens at a conservative boundary?

1 At a **constructive plate boundary**, two plates move apart from each other.

2 At a **destructive plate boundary**, two plates move towards each other.

3 At a **conservative plate boundary**, two plates slide past each other.

PRACTICE

1 Use the map to name a plate that is moving northwards.

2 Use the map to describe the distribution of earthquakes or volcanoes.

3 Name the two plates that form a constructive plate boundary.

Plate tectonics 2

THE BARE BONES

➤ At a constructive plate boundary new land is created.
➤ At a destructive plate boundary land is destroyed.
➤ At a conservative plate boundary land is neither created nor destroyed.

A Constructive plate boundaries

KEY FACTS

1 There are three main types of plate boundary.

2 At a constructive, or <u>divergent</u>, plate boundary, two plates <u>move apart from each</u>

- As the plates move apart, **magma** rises through the gap in the earth's crust and **cools** down to form **new crust**.

- An example of a **constructive** plate boundary is the **Mid-Atlantic Ridge**. Here, a **chain** of **underwater volcanoes** has **formed** along the plate boundary.

Q Can you describe what happens at a constructive plate boundary?

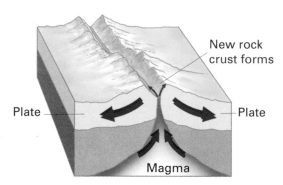

New rock crust forms

Plate — — Plate

Magma

B Destructive plate boundaries

KEY FACT

1 At a destructive, or <u>convergent</u>, plate boundary, two plates <u>move towards each other</u>.

2 When an oceanic plate moves towards a continental plate, the **denser oceanic plate** sinks **below** the **continental plate**.

- As the oceanic plate sinks, it **melts** and forms **magma**.

- Magma sometimes rises through cracks in the continental crust to form a volcano.

3 At a destructive plate boundary, the movement of the plates may push the continental crust upward to form **fold mountains**. This process is called **folding**.

Q Can you describe what happens at a destructive plate boundary?

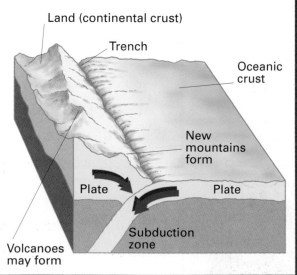

Land (continental crust)

Trench

Oceanic crust

New mountains form

Plate — — Plate

Subduction zone

Volcanoes may form

C Conservative plate boundaries

1 At a conservative, or <u>transform</u>, boundary, the plates <u>move horizontally past each other</u>, without creating or destroying the earth's crust.

Remember
People use different names for the different types of plate boundary. Find out which ones you are likely to see in your exam and learn these.

2 An example of a conservative boundary is the **San Andreas Fault**, USA.

Q Can you name a conservative plate boundary?

You could be asked to explain what happens at a particular type of plate boundary. Use a simple diagram to help your explanation (even if you are not asked to).

PRACTICE

1 Name a destructive plate boundary.

2 Using geographical terms, explain what happens at a destructive plate boundary.

3 Explain what is meant by the term 'folding'.

THE BARE BONES

➤ Earthquakes are caused when one of the earth's plates gets stuck. When the plates jolt free, pressure is released as waves of energy.

➤ Earthquakes tend to have a greater impact in Less Economically Developed Countries (LEDCs).

A What are earthquakes?

KEY FACTS

1 The movement of the earth's plates can cause earthquakes.

2 Plates do not move smoothly. Sometimes a <u>plate gets stuck</u>. <u>Pressure builds up</u> and when this <u>pressure is released</u> an earthquake can occur.

3 The point below the earth's surface where the pressure is released is called the **focus**.

4 The point on the earth's surface directly above the focus is called the **epicentre**.

5 **Shock waves** are produced when pressure is released from the focus. These waves are called **seismic waves**. The seismic waves are strongest at the epicentre of an earthquake. This is where the most damage is caused during an earthquake. The seismic waves spread out from the focus like ripples on a pond. As they travel outwards, they lose energy.

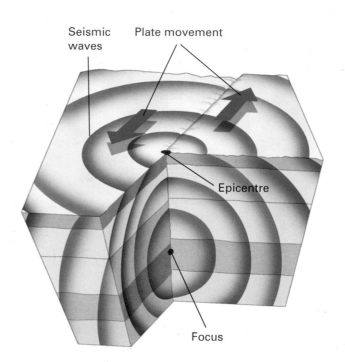

Remember
Earthquakes can occur at any type of plate boundary.

Q Can you explain what causes earthquakes?

B Measuring earthquakes

1 The strength of an earthquake is recorded using a machine called a **seismometer**.

• The seismometer picks up **vibrations and tremors** during an earthquake and produces a **seismograph** to show the strength of the earth's movements.

2 The **strength** or **magnitude** of an earthquake is measured using the **Richter Scale**.

Q How are earthquakes measured?

C The effects of an earthquake

1 Earthquakes can have a devastating effect on people and the environment.

- The **immediate** or **primary effects** of an earthquake include the collapse of buildings, roads and railways. People may be killed or injured and property damaged.
- The **long-term** or **secondary effects** of an earthquake include gas explosions and fires. Communications can fail, with telephone lines and computer links cut. Water can become contaminated as sewage and clean water pipes fracture. A lack of clean water can lead to the spread of disease.

2 Natural disasters such as earthquakes tend to have a <u>greater impact on LEDCs</u> than MEDCs.

- Buildings in LEDCs are not always strong enough to withstand the damage caused by earthquakes. LEDCs do not have sufficient healthcare facilities to deal with emergency situations. Access and communications in LEDCs tend to be poor. It is difficult to warn people about possible dangers or bring them emergency supplies. LEDCs also have limited money and resources to rebuild areas that are damaged.

D Preparing for earthquakes

1 It is hard to **predict** earthquakes. A **seismometer** can be used to **monitor tremors** inside the earth's crust and therefore identify potential earthquakes. However, earthquake predictions are **not accurate** enough to rely on. It is more worthwhile to invest money and resources in **preparing** for earthquakes.

2 There are many things people can do to **prepare** for an earthquake:

Training people to deal with an earthquake emergency (e.g. earthquake drills in schools).	Encouraging people to keep an earthquake kit at home, including first-aid items, tinned food and a radio.
Road and buildings can be constructed to reduce the damage caused by an earthquake (e.g. electronic shutters to cover windows).	Buildings can be constructed to be earthquake-proof. The building absorbs some of the energy released during the earthquake.

1 Explain what is meant by:
a) the focus of an earthquake   b) the epicentre of an earthquake.
2 Explain why the impact of an earthquake tends to be worse in LEDCs than in MEDCs.
3 With reference to a named example, describe the effects of an earthquake on the people living in an earthquake zone.

THE BARE BONES

➤ Volcanoes can form at both constructive and destructive plate boundaries.

➤ There are two main types of volcano: cone and shield volcanoes.

A The formation of volcanoes

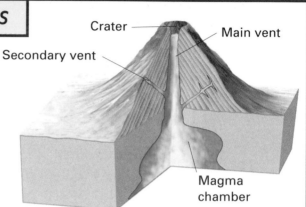

1 Volcanoes are formed when semi-molten rock called **magma** forces its way to the surface through weaknesses in the crust. As the magma rises, the surface pressure builds up within the earth causing a volcanic explosion.

Q What is lava?

2 When the magma reaches the surface of the earth, it becomes **lava**.

B Types of volcano

Volcanoes can be **classified** (grouped) in a number of different ways.

1 One way of classifying volcanoes is to look at their **shape** and **composition.**

Cone volcanoes – Tend to be found at destructive plate boundaries. **Tall** and **steep-sided.** Formed by **acid lava**, which is thick and **viscous** (sticky). The acid lava flows slowly and hardens quickly. This explains the steep-sided shape of these volcanoes. **Erupt violently** throwing out steam, gas and lava bombs.

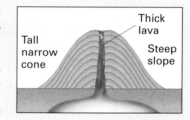

Shield volcanoes – Tend to be found at constructive plate boundaries. **Low** with **gentle slopes.** Formed by **basic lava**, which is thin and runny. Basic lava spreads quickly to form low, shield-shaped volcanoes. **Erupt frequently and gently.**

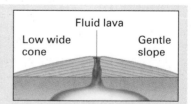

Some are composed (made up) of lava and ash, others are made up of lava only. **Composite volcanoes** are made up of alternating layers of lava and ash. Others are made from lava only.

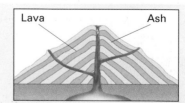

B

2 Another way of classifying volcanoes is to look at **how likely they are to erupt.**

Q Can you name the main features of a volcano?

Types of volcano

Active A volcano that has erupted recently and is considered likely to erupt again (e.g. the Nyragongo Volcano in the Democratic Republic of Congo, Africa – last erupted January 2002).

Extinct A volcano that is not considered likely to erupt again.

Dormant A volcano that has erupted within the last two thousand years but is not currently active (e.g. Mount Vesuvius in Italy).

C ## What are the effects of a volcanic eruption?

KEY FACT

Volcanic eruptions can have a <u>devastating effect</u> on people and the environment. <u>Lives can be lost and landscapes destroyed</u>.

The impact of a volcanic eruption

X Hot blasts of gas are sent into the atmosphere at the beginning of a volcanic eruption.

X Ash is thrown high into the atmosphere and covers the surrounding landscape as it settles. Sometimes ash remains in the atmosphere for several days and it can travel thousands of miles before settling on the earth's surface.

X When a volcano explodes, lava bombs can be thrown into the air. These are made up of pieces of rock and ash.

X Lava flows can destroy settlements (towns, cities and villages) and clear areas of natural vegetation, such as woodland.

X The heat of the volcanic eruption can melt snow and ice, causing fast-moving mud flows called **lahars**.

Q Make a list of the effects of a volcanic eruption.

D ## Why do people live in areas of volcanic activity?

Remember
Volcanoes and earthquakes do not happen only in LEDCs. North America is also affected by natural hazards.

Although volcanoes can be very dangerous, many people continue to live near them, often in very large numbers (high densities). Volcanoes can bring people many benefits:

✓ The **ash** deposited during a volcanic eruption adds **valuable nutrients** to the soils and helps to fertilise it. This helps **agriculture** and farming.

✓ Volcanic landscapes are very **attractive** and attract **tourists**. This brings in valuable income to the area.

✓ Many people believe that the **hot springs** created by volcanoes improve health.

✓ Although many people **choose** to live near volcanoes, **some people are too poor** to leave their home (particularly in LEDCs). Even if someone can afford to leave the area, they may be **attached to their home** and surroundings. Many people have to be forced to leave if a volcano becomes dangerous – they don't want to leave their home.

Q Can you list the advantages of living close to a volcano?

PRACTICE

1 Describe how a volcano is formed.

2 Using named examples, explain why many people live in areas of volcanic activity.

If you are writing about geographical processes in your exam, make sure you use the correct terms and refer to places you have studied.

Rocks and weathering

THE BARE BONES
- ➤ There are three groups of rocks: igneous, sedimentary and metamorphic.
- ➤ Rocks can be broken down by the process of weathering.
- ➤ Weathering and erosion help shape different landscapes and create distinct landforms.

A Different types of rock

KEY FACTS

1 The earth is made up of many different types of rock.

2 Rocks can be classified (organised) into three main groups.

Igneous rocks
These rocks are formed when **magma** from inside the earth **cools** and **solidifies**. Examples of igneous rocks include **granite** and **basalt**.

Sedimentary rocks
These rocks are formed from **fragments** of other rocks or the **remains of living things** that have been **compressed** into rocks. Examples of sedimentary rocks include **sandstone, chalk** and **limestone**.

Metamorphic rocks
These are **existing** rocks that have been **changed** by **intense heat** and **pressure** to form new rocks. The new rocks are harder and more compact than the original rocks. Examples of how metamorphic rocks can form include **limestone** becoming **marble** and **sandstone** becoming **quartize**.

Remember
A landscape can be made up of several different rock types.

Q Can you list the three main types of rocks?

B Different landscapes

KEY FACT

1 The study of rocks and different rock types is called geology. The geology of an area is one of the key factors influencing the shape of the landscape.

2 **Stronger rocks** tend to produce **highland** areas, whereas **weaker rocks** tend to form **lowlands**.

3 Whether rocks allow **water** to pass through them tells you whether they are **permeable** or **impermeable**.

• Water will pass through permeable rocks. Water is not able to pass through impermeable rocks.

4 The **permeability** of rocks will **determine** how **wet** or **dry** the surface of a landscape is.

• **Limestone** is a permeable rock that tends to form **dry upland** areas with few streams and thin soils.

• **Clay** is an impermeable rock that tends to produce **wet lowland** areas.

Q Name a type of permeable rock and a type of impermeable rock.

C Weathering

1 Weathering involves the <u>breakdown of rocks</u> on the earth's surface.

2 There are three types of weathering: **physical**, **chemical** and **biological**.

Physical weathering

This is caused by **changes** in **temperature** or **pressure**. There are two main types of physical weathering:

Remember
It is not just natural scenery that is shaped by weathering. Weathering and erosion can affect buildings, roads and pavements.

- **Freeze-thaw weathering** takes place when water is trapped in the cracks of a rock and freezes. The frozen water **expands** and **enlarges** the cracks. The process of freeze-thaw weathering repeats until the rock is **weakened** and **shatters**.

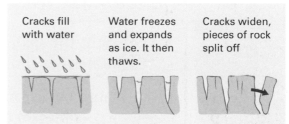

Cracks fill with water | Water freezes and expands as ice. It then thaws. | Cracks widen, pieces of rock split off

- **Onion-skin weathering** takes place in **hot desert climates**. During the heat of the day the surface of the rocks **heats up** and **expands**. At **night** it is **cold** and the rocks **contract**. This causes thin layers of rock to **peel off**.

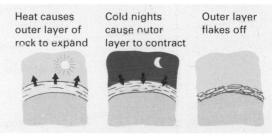

Heat causes outer layer of rock to expand | Cold nights cause outer layer to contract | Outer layer flakes off

Chemical weathering

This occurs when **weak acids** in **rainwater** attack and **break down** rock surfaces. Limestone is weathered in this way.

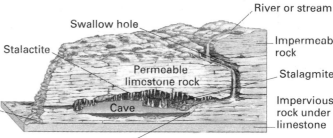

River or stream · Swallow hole · Stalactite · Impermeable rock · Permeable limestone rock · Spring · Cave · Stalagmite · Impervious rock under limestone · Stream runs underground as limestone disolves

Biological weathering

This involves **plants** and **animals**. Plant seeds can begin to grow in the cracks of a rock. As the **roots expand**, the cracks in the rock expand, the rock **weakens** and pieces break off.

3 Once rocks have been **broken down** they may be **eroded** or **transported**.

- **Erosion** is the **wearing away** of the land by **water**, **ice** or **wind**.

Q Describe how chemical weathering shapes limestone scenery.

1 What is the difference between erosion and weathering?

2 How is igneous rock formed?

3 Explain why there are few streams in limestone areas.

Glaciation

THE BARE BONES

➤ During the Ice Age, many areas across the world were permanently covered by ice, including the UK.

➤ A glacier is a body of ice that moves through a valley.

➤ Glaciers have shaped many landscapes through the processes of erosion, transportation and deposition.

A The Ice Age

KEY FACT

1 During the last Ice Age, 2 million to 10 000 years ago, the climate was much colder than it is today.

Remember
Glaciers are still shaping parts of the world today.

- In **upland areas**, **snow** remained on the ground **all year**. As more and more snow was added each year, it slowly **compressed** to form **ice**.

- In some areas, enormous **ice sheets** covered the whole landscape. The whole of **northern Britain** was **covered** in ice.

- In other places, ice only filled the valleys, forming **glaciers**. These glaciers **moved downhill** and **shaped the landscape**.

2 Glaciers shaped the landscape in many parts of Britain.

Q Name an area shaped by glaciation.

- **Today** ice and snow still **permanently** cover countries within the **Arctic Circle**, such as Greenland, Northern Canada and parts of Russia.

B Glacial erosion

1 Glaciers form in **hollows** on the **colder, sheltered** side of a mountain. These hollows are called **corries**.

2 Snow and ice gathers in the hollow and over time the corrie gets **larger** through **freeze-thaw** weathering (see p.25). Freeze-thaw weathering also **loosens** pieces of rocks, which then fall onto the glacier. This material is called **moraine**.

3 Inside the hollow, the ice begins to move in a **circular motion**, called **rotational slip**.

4 Eventually, the **ice** will **move out** of the corrie and over the **lip** of the hollow. The glacier moves down the mountainside. At the front of the glacier is the **snout**. A lake, called a **tarn**, may form in the corrie. The **steep knife-edged ridge** between two corries is called an **arête**.

5 As a glacier moves, it **erodes** the landscape in two main ways:

Q Can you explain how a corrie is formed?

Abrasion occurs when **pieces of rock** carried by the ice **wear** away the landscape. As the glacier moves, it transports **material** with it, which helps the glacier erode and shape the landscape. The material can be **frozen within** the glacier or **found underneath** it.

Plucking occurs when **meltwater** under a glacier **freezes** on to the **rock surface**. As the **glacier moves** forward, it pulls away **large fragments** of rock from the surface.

C Glacial deposition

1 As a glacier moves downhill, temperatures rise and the glacier begins to melt.

2 As the glacier **melts**, it **deposits** the material it has been carrying.

3 This deposition usually takes place in **lowland areas**.

4 <u>Glacial deposition</u> creates a number of distinct <u>landscape features</u> in lowland areas.

- **Moraine** is the rock material carried by the glacier. It is later deposited to form mounds of unsorted rocks and rock particles, which are called moraines. There are several different types of moraine. They are classified according to when and where they were deposited by the glacier.

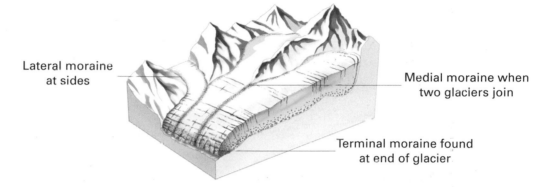

Lateral moraine at sides

Medial moraine when two glaciers join

Terminal moraine found at end of glacier

- **Drumlins** are mounds of boulder clay, deposited by glaciers and shaped by the moving ice. The ice moved over the drumlins to form small egg-shaped hills.

- **Erratics** are rocks transported many miles by a glacier and later found in an area of a different rock type.

5 **Glacial landforms** provide an important resource for the **tourist industry**.

- Glaciation helped to create distinctive and **interesting landscapes** such as the Lake District. **Tourists** and **day visitors** may visit these areas to go walking, hiking, mountain climbing or absailing.

- Many glaciated landscapes are protected from large-scale urban or industrial development. Many are **National Parks** in rural areas, where the main land use is for **primary industry**, such as farming or mining.

6 **Conflict** can sometimes occur between the different groups of people using National Parks. For example, tourists create a lot of **traffic**, which can cause **pollution** and inconvenience. Meanwhile, farmers want to protect their land from **trespassing** tourists and erosion by hikers.

Remember
You will need to know about features of glacial erosion and features of glacial deposition.

Q Draw a diagram to show the main features of glacial deposition.

PRACTICE

1 Using geographical terms, explain how a glacier erodes the landscape.

2 For a named glaciated environment that you have studied, describe how the land is used and explain why conflict may occur between the different groups of people using that landscape.

Practise drawing simple sketch diagrams to show glacial features.

The river system

THE BARE BONES
➤ The river system is part of the hydrological system (the water cycle).
➤ Water falling around a river may reach the river channel in a number of different ways.

A Systems in geography

KEY FACT

1 <u>Geography is often about systems</u>. For example, there are ecological systems, farming systems and industrial systems. Geographical systems can be <u>open</u> or <u>closed</u>.

- An **open** system has **inputs** and **outputs**.
- A **closed** system has **no inputs** or **outputs** – it is a **continuous cycle**.

2 A river is an **open** system with inputs, processes and outputs.

3 The river system is part of a larger system called the **hydrological cycle**.

Q What is an open system?

B The hydrological cycle

1 The hydrological cycle is a **closed system**. Water is continuously transferred from the world's **oceans** into the **atmosphere** and then to the **land**, before returning back to the **oceans**.

2 Water moves through the hydrological cycle via a series of **flows** or **transfers**. Water is also **stored** within the system; for example, in a lake.

Remember
Precipitation means any water that is released from clouds, including rain, sleet, snow and hail.

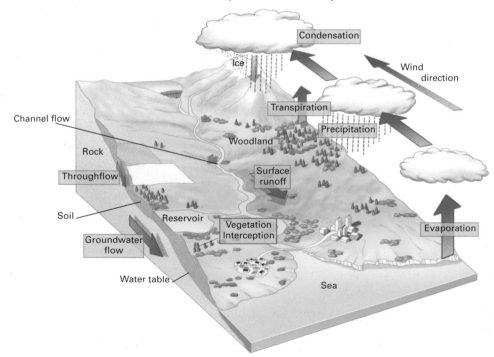

Q Can you list the ways in which water is transferred through the water cycle?

c   The river system

1   The river system is an <u>open system</u>, with inputs, processes and outputs.

2   The **source** is the point where a river begins its journey. The **mouth** is where the river reaches the end of its journey.

Remember
Drainage basins can be large or small. The River Amazon has one of the world's largest drainage basins.

3   The area of land **drained** by a river and its tributaries is called the **drainage basin** (sometimes called a **river basin**).

- **Tributaries** are the **small rivers** that **join** the **main river channel**. The point at which two rivers meet is called a **confluence**. The **greater the number** of tributaries, the **denser** the drainage basin.

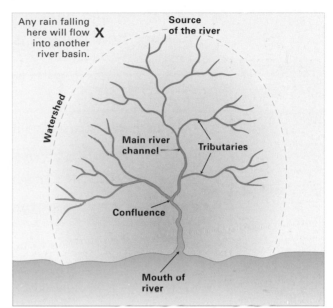

Any rain falling here will flow into another river basin. **X**

Source of the river

Watershed

Main river channel

Tributaries

Confluence

Mouth of river

4   The **boundary** between two drainage basins is known as the **watershed**. A watershed is usually a ridge of **highland**.

5   Water falling to the ground can travel to a river in many ways (some of these **flows** and **transfers** are similar to those found in the hydrological cycle):

- Water may **infiltrate** into the soil layer and move slowly towards the river. This is called **throughflow** (or soil flow).

- Water may **percolate** through the soil to the rock layer below. Water moving through the rocks towards the river is called **groundwater** flow or **base flow**.

- Some water flows directly over the ground to the river. This is called **overland flow** or **surface runoff**.

- Some water **does not reach** the river channel as it is **intercepted** by **vegetation** or is **stored** in the system (e.g. as a lake).

Q Can you name the main features of a drainage basin?

1   Explain what is meant by the following terms:
a) drainage basin
b) watershed

2   What is 'throughflow'?

3   Describe the ways in which water may reach a river channel.

Flooding and hydrographs

THE BARE BONES

➤ River discharge is the amount of water passing through a river at a given point.

➤ Hydrographs are used to show changes in river discharge over time.

➤ A significant rise in river discharge could lead to flooding.

➤ The natural and human environment surrounding a river can influence how likely it is to flood.

A River discharge

1 River discharge is measured in **cumecs** (cubic metres per second).

KEY FACT

2 The discharge of a river varies over time. Both <u>human</u> and <u>physical factors</u> influence the amount of water passing through a river.

Weather – physical	Geology – physical	Irrigation – human	Engineering – human
A period of **heavy rain** will increase river discharge. A period of **drought** will lower river discharge.	The rocks that make up the channel of a river may influence river discharge (e.g. in an area of **permeable rock**, the discharge may be lower than in an area of **impermeable rock**).	Sometimes water from a river may be **diverted** from the main channel and used by humans for **agriculture** or **industry**.	Humans can engineer or **change the natural course** of a river. Depending on what changes are made, the discharge may increase or decrease.

Q Can you explain what is meant by river discharge?

B Flooding

KEY FACT

1 <u>Severe weather and heavy precipitation</u> are the most common causes of river floods. The faster the water reaches a river, the more likely it is to flood.

Remember
Many people believe that there is a link between global warming and the rise in flooding across the world. This idea is open to debate; it is not something that everyone agrees upon.

2 A flood occurs when a river **overflows** its **banks**.

✗ Floods can be **devastating**, claiming lives and destroying homes.

✓ Floods can **benefit people** and the **environment**. For example, when a river floods, it deposits fine **silt** and **sediment**, which helps to **fertilise** the **soil** and generates excellent conditions for **farming**. People living near rivers such as the Nile in Egypt rely on regular flooding.

✗ Floods tend to have a **worse impact** on **LEDCs** than they do on **MEDCs**. An example of an LEDC that suffers from regular flooding is Bangladesh. LEDCs do not have enough equipment to accurately predict floods. Poor communications can make evacuation difficult and LEDCs lack the resources needed to rebuild their country once the flood waters subside.

B

3 The **landscape** around a river will influence how fast water reaches the main river channel.

- In **wooded areas**, trees may intercept precipitation. They may trap rainwater on their leaves and some water will be taken up their roots.

- In **rural areas**, fields that have crops or plants growing in them will intercept rainfall. However, if a field is left bare with dry clay soil, the water will not permeate into the soil and will travel overland to the river.

- **Steep slopes** in **highland areas** cause water to flow quickly downwards towards the river (**overland flow**).

- In **urban areas,** the landscape is made up mainly of **impermeable** rock (concrete). This means that water travels quickly towards the river by **overland flow. Drains** also take **water directly** to the river channel. Many houses in urban areas have **sloping roofs** which also increase runoff.

Q List those factors that make flooding most likely in a river.

C Hydrographs

1 Changes in river discharge over time are shown using a **hydrograph**.

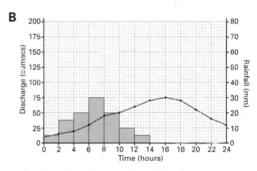

Remember
On a hydrograph rainfall is shown using bars, whilst discharge is shown using a line.

2 A hydrograph shows two variables: **rainfall** and **river discharge**.

3 The difference between the **peak** (highest) **rainfall** and the **peak discharge** is called the **lag time**.

4 The **longer** the lag time the **less chance** there is of a river **flooding**. A **short lag time** means that the water has reached the river channel quickly and there is a **higher risk of flooding**.

5 The **base flow** on a hydrograph shows the volume of water reaching a river through groundwater flow.

- The **rising limb** shows the **increase** in river discharge, while the **falling limb** shows the **decrease** in river discharge.

Q Can you calculate the lag time for each of the graphs shown opposite?

PRACTICE

1 Give three reasons why river discharge may vary between two different river basins.

2 At what time was the peak discharge on graph B?

3 Which graph, A or B, shows river discharge patterns for an urban area? Explain your answer.

If you are given a hydrograph in an exam, make sure you read the axis carefully.

Flood management

THE BARE BONES
➤ Flooding can be reduced through flood-management schemes.
➤ Flood management can involve hard engineering schemes or natural flood-control schemes.
➤ Flood management has both advantages and disadvantages for people and the environment.

A Flood management

KEY FACT ▶

1 The <u>impact</u> of flooding <u>can be reduced</u> in a variety of ways.

2 **Flood-protection measures** can be put in place to **control river discharge, predict** possible flooding and **reduce** the impacts of floods on people and the environment.

- **Hard options** tend to involve making significant changes to the natural river channel. These changes tend to be costly and long-lasting.
- **Natural options** tend to be low-cost and do not involve making significant changes to the natural river channel.

3 **Flood management** and **prediction** tends to be more effective and accurate in **MEDCs**. LEDCs often lack the necessary resources for effective flood management.

Q Can you explain the difference between hard engineering and natural flood control?

B Hard engineering options

KEY FACT ▶

1 Hard engineering options tend to be <u>costly</u> and have a <u>significant impact</u> on the <u>natural environment</u>.

2 **Building a dam**: Constructing a dam in the upper part of a river allows people to control the amount of discharge in the river further down its course.
- ✓ Water is held back behind the dam and can be released in controlled amounts.
- ✓ The water is usually stored in a reservoir. It can then be used to help generate power (hydroelectric power) or for leisure and recreation, such as water sports.
- ✗ Building dams is expensive and can spoil the look of the natural environment.
- ✗ A dam may cause sediment to get trapped in the upper sections of a river, resulting in coastal erosion at the mouth of the river.
- ✗ As the river is no longer able to flood naturally, the fertility of the flood plain may be reduced. This can affect agriculture, particularly in LEDCs. This has happened along the River Nile due to the Aswan Dam.

3 **Modifying the river channel**: The river channel can be straightened and deepened.
- ✓ Changing the river channel can allow a larger amount of water to flow quickly through the river.
- ✓ The water flows away from risk areas quickly.
- ✗ There is an increased risk of flooding downstream as the water reaches these areas faster.

Remember
Many LEDCs have borrowed money to build dams and other flood-control methods. This has created problems of debt for some LEDCs.

Q Make a list of the benefits and problems of building a dam?

C Soft engineering options

1 Soft engineering options tend to be <u>low–cost</u> and <u>do not have a significant impact upon the natural environment</u>. Soft options are also easier to maintain than hard engineering options.

Remember
Soft engineering can often be more effective than hard engineering.

2 Afforestation: This involves planting trees and vegetation around the river channel, helping to lower river discharge and therefore reduce the risk of flooding. The leaves and roots of trees intercept rainwater and reduce runoff. Trees such as willow are often chosen for afforestation schemes as they take up a lot of water.

Q What is afforestation?

3 Ecological flooding: Some parts of the river are allowed to flood naturally in rural areas, to prevent flooding in urban areas.

4 Land use planning: Governments and local authorities can prevent or limit the number of homes constructed in areas where there is a risk of flooding.

D Case study – the River Rhone

Remember
Many modern flood-management schemes use a mixture of hard and soft engineering options.

1 The River Rhone travels from its source high up in the **Swiss Alps**, through France to its mouth in the **Mediterranean**.

2 At its source, the River Rhone enters **Lake Geneva** and travels through the city of Geneva to **Lyon**, in France. In the past, the river frequently flooded here, so engineers have tried to change the river's course. **Floodgates** have been added to control the amount of water passing downstream. New **canals** have been built alongside the river to allow shipping to pass more easily. Where this has happened, the old course of the river is left to take in excess water during flooding.

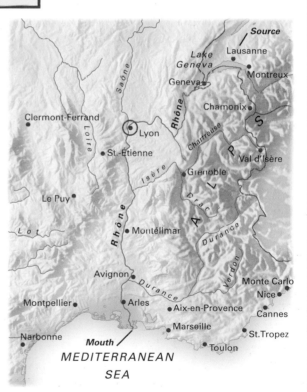

Q Can you describe three methods of natural flood control?

PRACTICE

1 Give two reasons why humans may want to control flooding.

2 Using examples, describe how people try to control river flooding.

3 For a named flood-management scheme that you have studied, explain the advantages and disadvantages of the scheme for people and the environment.

River processes

THE BARE BONES

➤ A river contains energy. It uses this energy to carry out river processes and shape the landscape.

➤ There are three main river processes: <u>erosion</u>, <u>transportation</u> and <u>deposition</u>.

➤ The journey of a river from source to mouth is called the long profile.

A The long profile of a river

1 A river changes as it travels from its **source** to the **mouth**.

2 The course of a river can be divided into three main sections: the **upper course**, **middle course** and **lower course**.

3 Rivers begin flowing in **highland** areas and flow downwards to **lowland** areas.

Q Where is the source of a river?

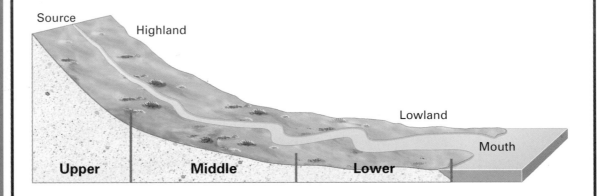

B Erosion

KEY FACT

1 Erosion is the <u>wearing away</u> of the <u>riverbed</u> and <u>banks</u>. It can also involve the wearing away of <u>rocks and particles</u> being carried by the river.

2 Rivers can erode in **four** ways:

- **Hydraulic action:** The force of the water wears away the bed and banks of the river.

- **Abrasion (or corrasion):** Rocks and pebbles being carried by the river wear away the bed and banks. Sometimes this material gets trapped in a dip, where it swirls around to form a pothole.

- **Attrition:** Rocks and pebbles being carried by the river knock together and are broken down into smaller particles.

- **Solution (or corrosion):** The water dissolves rocks and minerals in the river channel. Chalk and limestone are two rock types that dissolve relatively easily.

Remember
Some of the methods of river erosion sound similar. Each of them is different, so learn them carefully.

B

3 Erosion can be **vertical** or **lateral**.

Vertical or downward erosion takes place in the **upper course** of the river, as **gravity pulls** the **water downwards**. Vertical erosion **deepens** the river channel and can create **V–shaped valleys**.

Lateral (sideways) erosion takes place in the **lower** and **middle course** of the river. Lateral erosion **widens** the river channel.

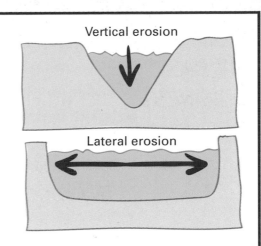

Q Can you list the four methods of erosion within a river?

C Transportation

Rivers can use their energy to **transport** material.

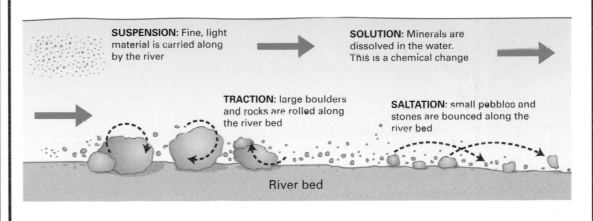

Remember
The size of the material being carried by the river affects the way it is transported.

Q Can you describe the different methods of transportation within a river?

D Deposition

KEY FACT

1 If a river loses energy and slows down, it <u>deposits (drops)</u> the load that it has been carrying.

2 A river may lose energy if:
- it enters an area of shallow water
- it enters an area of vegetation
- the volume (amount) of water in the river decreases (e.g. after a flood).

3 Most deposition takes place in the **lower course** of the river.

Q Why may a river deposit its load?

PRACTICE

1 Explain the difference between abrasion and attrition.

2 How do you think the size of particles carried by a river will change as you move from source to mouth? Explain your answer.

THE BARE BONES

➤ Rivers travel from their highland source to the mouth of the river, found in a lowland area.

➤ In the upper course of the river, vertical erosion can create V-shaped valleys and interlocking spurs.

➤ In the lower course of the river, lateral erosion can help to create meanders and sometimes ox-bow lakes.

➤ Close to the mouth of the river, a wide flood plain, and sometimes a delta, is created.

A Features in the upper course of a river

KEY FACT

1 In the upper course of a river, the <u>gradient</u> of the landscape is <u>steep</u>.

2 At its source, a river erodes vertically (downwards), cutting into the landscape to form a **steep-sided, V-shaped valley**.

3 In the upper course of the river, the channel is **narrow** and the river **erodes** its way through areas of **softer rock**. As the river **cuts through** the **softer rock**, it **winds** to **avoid** areas of **hard rock**. This can lead to the formation of **interlocking spurs**.

4 **Rapids** may also form in the upper course of the river, when the river travels over alternating bands or hard and soft rock.

5 **Waterfalls** are a common feature in the upper course of many larger rivers. A waterfall occurs when a **layer of hard resistant rock** lies **over a layer of softer rock**, which will erode more easily:

Remember
In the upper course, the river is youthful, which means that the water is clear and appears to flow very fast. The flow will actually get faster as the river moves down stream and more water is added to the channel.

Q Name three features that you may find in the upper course of a river.

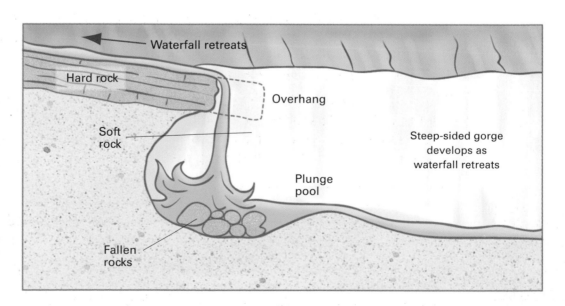

Waterfall retreats

Hard rock

Overhang

Soft rock

Steep-sided gorge develops as waterfall retreats

Plunge pool

Fallen rocks

B Features in the lower course of a river

1 In the lower course of the river, the gradient is gentler than in the upper course. The river has <u>more energy</u> and the <u>volume</u> of water is high.

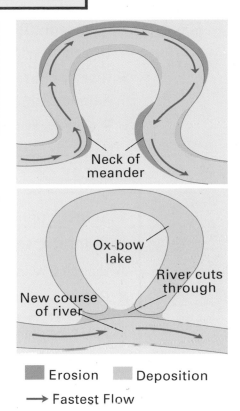

Neck of meander

Remember
An ox-bow lake is different to a meander and takes many years to form.

2 In the lower course there is more **lateral** (sideways) **erosion**. The channel is **wide** and **deep**. The river has less **friction** to overcome, which means that the river can flow **faster**.

3 As the river erodes sideways, it swings from side to side, forming large bends called **meanders**.

Ox-bow lake

River cuts through

New course of river

4 Over time, the loop of a meander becomes tighter. If it **becomes** too **tight** the river may simply **cut across** the **neck** of the meander to form a **straight** river channel. The loop is **cut off** from the main channel and forms an **ox-bow lake**.

Q Can you explain how a meander is formed?

■ Erosion □ Deposition
→ Fastest Flow

C Features at the mouth of a river

1 As the river reaches the mouth, it has a <u>large discharge</u> and the river channel is <u>deep</u> and <u>wide</u>.

• The valley is now **wide** and **flat**. This creates a wide **flood plain** around the river.

2 A flood plain is a flat area around a river that regularly floods. Each time a river floods, **silt (alluvium)** is **deposited** on the flood plain. This makes it very fertile and good for farming and agriculture. Flood plains are often highly populated, particularly in LEDCs, where farming employs many people.

Q Can you list three differences between the upper and lower course of a river?

3 **Deltas** are a feature found at the mouth of large rivers (e.g. the Ganges Delta). When a river enters the sea, it **deposits** its load; if this occurs **faster** than the sea can remove the material, a **delta** may form. Over time, it becomes a permanent land feature, rich in **alluvium** and providing **fertile farmland**.

PRACTICE

1 Using a simple diagram, explain how an ox-bow lake is formed.

2 With reference to a named example, explain why people often live in high densities (high numbers) in the lower course of a river.

Take coloured pencils into the exam to make quick sketches of river features. Colour will help you to highlight important aspects.

River pollution

THE BARE BONES

➤ River pollution is caused when humans discharge their waste into the river system.

➤ River pollution can harm human health, destroy wildlife and damage the natural environment.

➤ Rivers carry pollution to the seas and oceans.

A Sources of river pollution

Remember
Pollution also takes place in seas and oceans. A river may carry the pollution to the sea or waste may be dumped directly into the sea.

1 Sources of river pollution can be **classified** (organised) in a number of ways:

• **Domestic sewage**

✓ Drains carry sewage from people's homes to **sewage works**, where it is treated.

✗ However, sometimes **raw** (untreated) **sewage** may **leak** from treatment works and enter a river basin. In a flood, sewage may overflow from the drains and enter rivers as **surface runoff**.

• **Agricultural waste**

✓ Farmers often use **chemicals** to help their crops grow and survive.

✗ These chemicals (**herbicides** and **pesticides**) can enter river systems through the soil (throughflow) or rocks (groundwater flow). **Slurry** (animal waste) is also stored on farms. This can leak through the soil to nearby rivers.

• **Industrial waste**

✗ **Factories** and **businesses** produce waste and sewage. Some factories discharge their waste into rivers (either by accident or intentionally). Many factories use water for **cooling** during the **production** process. Often this **warm water** is discharged directly into rivers.

• **Waste produced from energy generation**

✗ Power stations use large amounts of water for cooling. This water **warms up** during the cooling process and is then often discharged into rivers. Power stations can also be responsible for causing acid rain, which then pollutes rivers and lakes.

Q Can you list the three main sources of river pollution?

B Impacts of river pollution

KEY FACT

1 River pollution can affect <u>people</u>, <u>wildlife</u> and the <u>environment</u>.

2 Loss of oxygen: River creatures and river plants need oxygen to survive. When raw (untreated) sewage enters a river system it is broken down by bacteria in the water to create ammonia. However, this process uses up a great deal of oxygen. A lack of oxygen can destroy river wildlife.

B

Q Can you explain how river pollution can harm river wildlife?

3 Eutrophication: This is the term for an increase in nitrate levels in a river, which can rise due to pollution. Too many nitrates (often from agricultural products) encourage plant growth, particularly algae. **Algae** uses up oxygen and blocks out light. This affects the amount of wildlife in the river and turns the water green.

4 Human health: Can be damaged by high nitrate levels in a river. In recent years high nitrate levels have been detected in drinking water in the UK and other parts of Europe. Reducing nitrate levels requires expensive water treatment processes.

C Case study – river pollution along the River Rhine

Remember
Don't get the River Rhine muddled up with the River Rhone. Both rivers are in Europe but the Rhine flows north towards the North Sea whilst the Rhone flows south towards the Mediterranean. The River Rhone features in a case study on page 33.

1 The Rhine is 1320km long and is one of Europe's major rivers. The Rhine runs from its **source** in **Switzerland**, through **France**, **Luxembourg** and **Germany** to its mouth in the **Netherlands**.

2 The River Rhine provides 20 million people with drinking water. It is also used for **transport**, **extracting water** and for **discharging waste**. Some of Europe's most industrialised areas are situated along the Rhine.

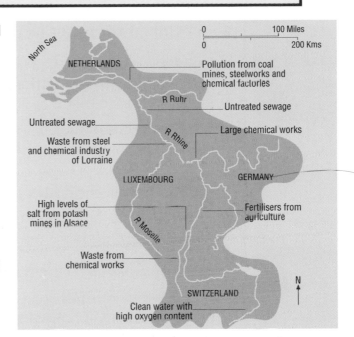

3 One major **problem** of managing pollution along the Rhine is that the river runs through many different countries. This makes it hard to agree on **who is responsible** for dealing with pollution issues along the river. The **Netherlands**, near the mouth of the Rhine, **receives the most polluted water**.

4 In the 1950s, the countries that use the Rhine formed the **International Commission for the Protection of the Rhine (ICPR)**. They have developed **schemes** to **reduce pollution** and **improve water quality**. They also **decide how much each country pays to the Commission**.

5 In recent years there has been evidence that **oxygen levels** in the Rhine are **rising** and the variety of **river wildlife is increasing**.

Q Can you explain why it is difficult to manage the River Rhine?

PRACTICE

1 What are the impacts of water pollution in rivers?
2 Using the map above, describe three sources of pollution.
3 With reference to a named example, explain how river pollution can be reduced.

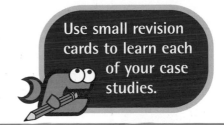
Use small revision cards to learn each of your case studies.

Water management

THE BARE BONES

➤ There is a growing demand for water in the UK.

➤ In the UK, most people live in areas with limited water supplies. The areas with good supplies of water are sparsely populated.

➤ Water supplies can be managed in a number of ways. Water management can help prevent shortages and control pollution.

A Water demand and supply in the UK

KEY FACT

1 **Demand** for water in the UK is **rising**.

- **Homes** and **industries rely** on a **constant supply** of **fresh water**. Many **leisure** and **recreation activities** also rely on a regular supply of water (e.g. gardening and water sports).

- **Modern technology** often demands a good supply of water (e.g. washing machines, dishwashers and electric showers).

2 Water supplies are not always **reliable** and we have begun to suffer from **regular water shortages** in the UK.

KEY FACT

3 One of the biggest problems in the UK is that **water supplies are not evenly distributed**.

- Many people live in the driest parts of the UK.

- The **heaviest rainfall** occurs in the **north** and **west** of the UK. However, **large numbers** of people **live in the south-east**.

Remember
As more and more countries across the world industrialise, the global demand for water increases.

Q Can you explain why some parts of the UK sometimes suffer water shortages?

Groundwater supplies

Surface supplies

Northumbrian

Yorkshire

North West

Severn Trent

Anglian

Welsh

Thames

Wessex

Southern

South West

B Water management in the UK

1 Managing water supplies in the UK involves ensuring that homes and businesses are provided with a <u>reliable supply of fresh water</u>, and also <u>controlling</u> levels <u>of water pollution</u>.

Remember
Many countries have built dams and reservoirs so they can manage water supplies and control flooding at the same time.

2 In England and Wales, water is **managed** by localised water authorities.

3 There are **ten water authorities** in England and Wales and each is based around a **major river basin**. They are responsible for:

collecting and distributing water supplies

treating and disposing of waste water from homes and factories

monitoring and controlling pollution levels in rivers

building and maintaining dams and reservoirs

regulating the use of rivers and reservoirs for industry and recreation (e.g. issuing fishing licenses).

4 The water authorities have taken a number of different measures to **manage water supplies** and **prevent water shortages**:

- **New reservoirs** have been built to **collect** and **store water**, such as Kielder Water in Northumberland.

- **Water transfers** have taken place, with water being **piped from wet areas** to drier areas.

- **Advertising campaigns** have been shown on television and radio to encourage people **not to waste water**.

- During the summer months some water authorities have imposed **hosepipe bans** in dry areas. This affects people who use water in their gardens and for cleaning their cars. It also affects public areas such as parks and golf courses.

Q Can you make a list of possible water management techniques?

PRACTICE

1 According to the map opposite, which water authority has the greatest supply of water?

2 With reference to named examples, describe how people can manage water supplies and prevent water shortages.

In your exam, you'll probably be asked to describe the pattern shown on a map. When you describe it, use your knowledge of directions (e.g. in the north or in the west). Also, quote place names and regions.

The sea at work

THE BARE BONES

➤ Coastlines are constantly changing due to the processes of erosion, transportation and deposition.

➤ The power of the waves determines the type of processes that take place.

A Wave energy

KEY FACTS

1 Waves are created when <u>wind</u> blows over the surface of the sea <u>creating friction</u>.

2 The size and energy of a wave varies according to its <u>speed</u> and the amount of <u>time</u> the wave has been <u>moving</u>.

Remember
The power of the waves depends on wind and the fetch, and determines the processes that take place.

- Waves are biggest and have the most energy when the **wind is strong**, has been blowing for a long time and has come a long way.

- The distance over which a wave has travelled is called the **fetch**.

3 Friction with the seabed slows down the bottom of the wave, but the top of the wave continues moving at the same rate and topples forward, breaking against a cliff or beach.

Swash is water that washes **up** a beach.

Backwash is water that washes back **down** a beach.

Water particles move in circular pattern.

Friction with seabed increases as water becomes shallow. Pattern becomes egg-shaped.

Top of wave not affected by friction.

Wave becomes steeper and breaks up the beach.

Wave direction

Sea bed

Q Can you define 'swash' and 'backwash'?

KEY FACT

4 There are two types of wave: <u>constructive</u> and <u>destructive</u>.

B Coastal processes

There are three main processes at work in the sea: erosion, transportation and deposition.

KEY FACT

1 <u>Erosion</u> is destructive waves wearing away the coastline. It happens when the waves are packed with energy. Erosion destroys landforms.

B

Waves erode in **four** ways:

- **Hydraulic action** is when waves crash against cliffs, trapping and compressing air and water in rock cracks. As the waves move back, pressure is released, causing the air and water to expand. This explodes, breaking off rock fragments.

- **Abrasion** is when breaking waves pound rocks and pebbles against the cliffs, wearing the land away in a sandpaper effect.

- **Attrition** is when waves smash rock fragments against each other, making pebbles smoother, rounder and smaller. In time, the particles are ground into grit and sand.

- **Corrosion** is when the chemicals in sea water dissolve or rot rocks such as limestone and chalk.

2 Transportation **is the movement of material in the sea and along the coast by waves.**

- Transport along the coast is when waves move material across a beach. This is called **longshore drift.**
- The prevailing wind causes the waves to break on the beach at an angle.
- Swash carries the material up the beach at an angle. Backwash drags the material back down the beach at right angles.
- Each wave pushes material further along the beach.
- Material is moved along the beach in the direction of the prevailing wind.

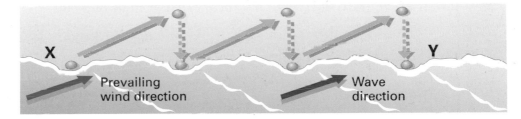

X   Prevailing   Wave
     wind direction   direction   Y

3 Deposition **is the dumping of eroded material on the land by constructive waves. It happens when the waves have less energy. Deposition creates landforms.**

PRACTICE

1 What causes waves?

2 Which factors control the amount of energy a wave has?

3 Explain the term 'fetch'.

4 Using the longshore drift diagram above, explain how material may be moved between points X and Y.

THE BARE BONES ➤ The processes of erosion, transport and deposition create coastal landforms.

A Rates of erosion

The rate at which any landform is created depends on:

Weathering
Sub-aerial attack on cliffs speeds up erosion.

Human activity above cliffs

Rates of erosion are affected by:

Wave energy
Greater when waves are big and frequent.

Rock type
Resistant rocks such as granite erode very slowly. Less resistant rocks such as clays erode very quickly.

Beaches
Beaches in front of cliffs absorb wave energy, slowing down erosion.

Rock structure
Cliffs with many weaknesses, called joints and cracks, are more easily eroded.

Q What factors affect erosion?

B Features of erosion

1 Making cliffs and wave-cut platforms.

• A **wave-cut platform** may be created in front of a cliff.

A cliff made from <u>soft rock</u> will erode quickly and form a <u>gently sloping</u> cliff. <u>Hard rock</u> creates <u>steep</u> cliffs.

KEY FACT

Remember
Soft rock erodes quickly, giving gentle landforms; hard rock erodes slowly, giving steep landforms.

Hydraulic action, abrasion and corrosion attack the coastline between the high and low water marks. This creates a **notch** at the base of the cliff and an **overhang** above.

The cliff above collapses.

Over time, the collapsed material erodes to form beach material and the cliff retreats, leaving a wave-cut platform in front of the cliff; for example, at Southerndown, South Wales).

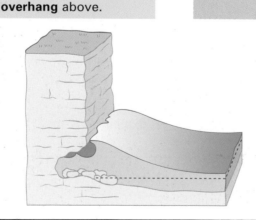

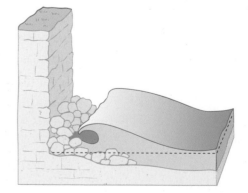

B

2 Making **headlands**.

> <u>Headlands</u> are found where there are different rock types at an angle to the sea (called a <u>discordant</u> coastline).

- Where there is softer rock, the rate of erosion will be fast, forming gently sloping **bays**. Where the rock is hard, erosion will be slower, forming a **headland** with steep cliffs that jut out into the sea (e.g. at Old Harry Rocks, Studland).

3 Making **caves**, **arches**, **stacks** and **stumps**.

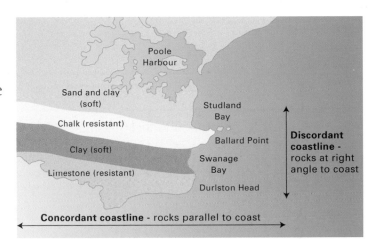

Poole Harbour

Sand and clay (soft)

Chalk (resistant)

Clay (soft)

Limestone (resistant)

Studland Bay

Ballard Point

Swanage Bay

Durlston Head

Discordant coastline - rocks at right angle to coast

Concordant coastline - rocks parallel to coast

> Caves, arches, stacks and stumps are made when a <u>narrow</u> <u>headland</u> made from <u>hard rock</u> is <u>eroded</u>.

Strong waves attack cracks and weaknesses in the rock by hydraulic action, abrasion and corrosion to make caves. (The cave may develop a blow hole if hydraulic action happens along a weakness in the cave roof, causing collapse.)

→ Arches form as the caves erode even more and break through the headland.

→ Stacks form when the roof of the arch collapses.

→ Stumps, which are only seen at low tide, are left behind once the stack has eroded.

Q Can you explain how each feature is formed in stages?

THE BARE BONES
➤ The processes of erosion may be speeded up or slowed down by geology, weathering and human activity.

A Features of deposition

Remember
The British coastline is 9500km long and gets some of the most powerful waves in Europe. It is not surprising that the sea has a massive impact on our lives.

Two main landforms are created: **beaches** and **spits**.

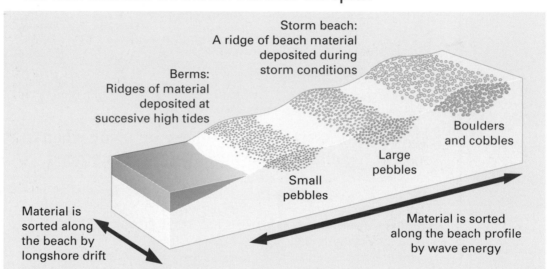

Storm beach:
A ridge of beach material deposited during storm conditions

Berms:
Ridges of material deposited at succesive high tides

Boulders and cobbles

Large pebbles

Small pebbles

Material is sorted along the beach by longshore drift

Material is sorted along the beach profile by wave energy

KEY FACT
1 <u>Beaches</u> are found where waves have transported and deposited eroded material from the sea.

• This can be on wave-cut platforms or in bays. Whether the beach is sand or pebble will depend on rock type and wave energy, as will the size and height of the beach.

• The beach profile above shows that the largest material is deposited on the **storm beach** closest to the cliff during storms. Below this are **berms**. These are ridges of material dumped at high tide. The smallest material is deposited closest to the shoreline.

KEY FACT
2 <u>Spits</u> are extended beaches of pebbles or sand that are joined to the land at one end and stretch out into the sea at the other end.

• They only form when certain conditions are present:

The sea is shallow and the current is weak, to allow deposition.

There is a sudden change in the direction of the coastline or a river mouth.

There is a prevailing wind and longshore drift.

A

Step 1: **Longshore drift** moves material beyond the change in coastline.

Step 2: The **spit** is formed when the material is deposited.

Step 3: Over time, the spit grows in length and may develop a **hook** if wind direction changes further out.

Step 4: Waves cannot get behind the spit, creating a sheltered area where silt is deposited and mud flats or **salt marshes** form.

Spurn Head, Humberside

Q Can you give an example of a spit?

PRACTICE

1 In which coastal feature would you find a 'notch'?

2 With which rock types are headlands associated?

3 Look at the diagram opposite. Name features A, B, C and D. Explain how D is formed.

4 Draw a labelled diagram to show how a spit is made.

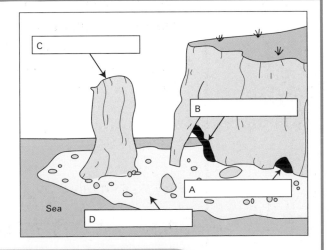

C

B

A

Sea

D

Put the diagrams you draw in an exam into a frame as in Question 3. Frames make your work easier to read

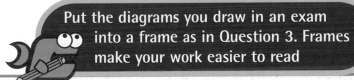

Coasts and people

➤ There are conflicts between managing coastal erosion to protect the environment and meeting the needs of groups of people.

➤ Both hard and soft engineering techniques can be used to protect coastlines.

➤ Protecting the coastline in one area can cause problems elsewhere.

A Why protect coastlines?

KEY FACT

1 Many coastlines in MEDCs are heavily populated.

2 Coasts provide people with **income** in the **tourism** and leisure industries.

• Many coastlines have **high economic and land value**.

Q Can you list two reasons for protecting coastlines?

3 Coastlines are prone to **flooding** or **erosion**.

4 Coastlines are **fragile** natural environments which are easily damaged by people.

• If they are destroyed, **ecosystems** take a long time to recover.

B Defending the coast

KEY FACT

1 The main strategies to defend against flooding and erosion use <u>hard and soft engineering techniques</u>.

• **Hard** approaches are expensive, short term, ugly and not sustainable as they battle against natural processes, sometimes causing damage in other places down the coast.

Remember
Coastlines in built-up areas of the UK are heavily managed.

• **Soft** approaches are less expensive, longer term, attractive and sustainable as they work with natural processes, causing less damage.

2 Hard management techniques:

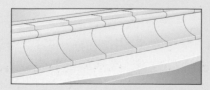

Groynes: Wooden barriers at right angles to the beach. Beach material builds up against the barriers, creating a wider beach that absorbs wave energy, slowing cliff erosion. Without groynes, this material is transported away by longshore drift.

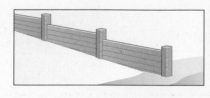

Sea walls: Curved concrete walls built at the base of cliffs or to protect a settlement from erosion. The waves are reflected back off the wall. Sometimes this wave energy depletes the beach. Over time the wall is also eroded.

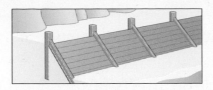

Revetments: Wooden slatted barriers built towards the back of the beach to protect the cliff. Waves break against the revetments, which dissipate the energy. The cliff base is protected by the beach material held behind the barriers.

Rock armour/boulder barriers: Large rocks are piled up in areas prone to erosion to absorb wave energy and hold beach material in place.

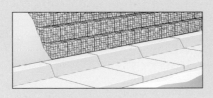

Gabions: Boulders and rocks are wired into mesh cages and placed in front of areas vulnerable to erosion. Wave energy is absorbed by the rocks, limiting erosion.

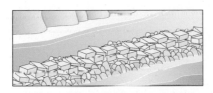

Offshore breakwater: Huge concrete blocks and natural boulders are sunk offshore to alter wave direction and limit longshore drift. This leads to wider beaches that absorb the reduced wave energy, protecting cliffs and settlements behind.

Cliff stabilisation: Prevents cliff slump by draining off excess rainwater to reduce water-logging. Terracing, planting and wiring hold the cliffs in place.

3 Soft management techniques:

Beach nourishment: Replaces natural beach material that has been removed by longshore drift with material dredged from the seabed. This creates a natural barrier that slows erosion by absorbing wave energy and provides a flood barrier.

Stabilising sand dunes: By planting grasses to hold dunes in place and introducing footpaths to reduce trampling, the rate of erosion is slowed and an effective flood barrier is created.

Managed retreat: Happens when the land adjacent to the sea is of low value. A decision is made to allow the land to erode and flood. Once this has happened natural defences such as beaches and salt marshes will form.

Q Can you list the main coastal management strategies?

PRACTICE

1 How do 'groynes' work?

2 What is beach nourishment?

3 What is the difference between hard and soft management strategies?

4 Why are some people against using hard management?

THE BARE BONES

THE BARE BONES

➤ A Cliff Protection Scheme at Mappleton was built in 1991 to save the village and the B1242.

➤ It has created multiple knock-on effects further down the coast.

A Facts about the area

1 Mappleton is a village on the east coast of Yorkshire, along a 50km stretch of coast called **Holderness**. The closest large settlement is **Scarborough**.

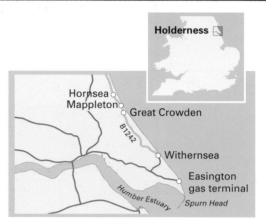

KEY FACT

2 The whole coastline is one of the most <u>vulnerable</u> in the world, retreating at a rate of one to two metres per year.

Q Where is Mappleton?

B The Cliff Protection Scheme and its impact

1 There are many reasons for the **fast erosion** along this coast:

Remember
You should be able to draw simple diagrams to explain your answer.

Cliffs slumping when water-logged.

Causes of erosion

Strong prevailing winds from the north-east and long fetch across the North Sea create powerful destructive waves.

Cliffs made of soft clay (glacial till).

Thin beach removed by longshore drift, exposing cliffs to full power of waves.

2 Some people in Mappleton felt it should be protected because:

• villagers were about to **lose** their **homes** and **businesses**
• the **main road** connecting towns along the coast (the B1242) was within 500 metres of the cliffs
• it would be very **expensive to rehouse** the people **and reroute** the road.

3 In 1991, it was decided that protecting Mappleton would benefit more people than it would disadvantage. A coastal protection plan costing £2 million was implemented that comprised two **rock barriers** built out into the sea north and south of the village to trap material and build up the beach. In addition, **rock armour** was placed along the cliff base.

• All these measures **absorb wave energy** and **reduce wave attack**, protecting the cliffs and village above.

4 By 1994, erosion had stopped at Mappleton. However, south of Mappleton, **longshore drift** is continuing to shift material towards **Spurn Head**, leaving little behind to protect the soft cliffs. The rate of erosion here has increased to 10 metres per year.

Q Can you say why Mappleton has caused such a problem?

B

Remember
Coastal protection is both a local and a national issue.

Q Who is Sue Earle?

- Some farmers are now being **forced to move** and lose their livelihoods. Sue Earle, owner of Clifftop Farm, moved out in 1996, when the cliffs came within 4 metres of her home. She has sued the local council for compensation.

5 There is concern at Spurn Head that the natural systems have been disrupted, affecting **unique habitats**.

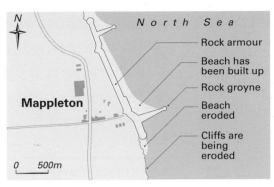

North Sea
- Rock armour
- Beach has been built up
- Rock groyne
- Beach eroded
- Cliffs are being eroded

Mappleton

0   500m

C   Issues to consider

KEY FACT

Q Can you relate to other people's values and attitudes?

Coastal protection is usually found where <u>land has economic value</u>.

Schemes like the one at Mappleton always create conflict between different groups of people. The dilemma is how to **resolve this conflict** for the best.

Reasons for favouring coastal protection
- Residents and entrepreneurs want to protect their homes and businesses
- Beaches are held in place, attracting tourists

Reasons for opposing coastal protection
- It can have disastrous impacts elsewhere
- It is expensive and interferes with nature
- It can destroy rare wildlife and habitats

PRACTICE

1 List the three main causes of erosion along the coast.

2 How may human action at one place cause a different action at another place?

> **Councillor Barbara Baughan** – If we look at the 1000m that we're talking about here, that's a £6 million scheme. The local authority would have to finance the whole scheme. We just cannot afford it. Whether it's this local authority or other local authorities. That sort of financing on one scheme couldn't be justified currently or perhaps in the future.

> **John Pethick** – The land here is not worth a great deal. It's agricultural land. Where the money's going is to protect the mud flats in front of some very important towns, such as Amsterdam, London and King's Lynn. An enormous amount of money is at stake. What is most important to us is what we ought to be asking. Should we be protecting the agricultural land here by stopping the erosion, or allowing the erosion to take place and protect some of the biggest cities in Northern Europe?

> **Professor Keith Clayton** – The only safe defence against the sea is a good beach. And you will never have any good beaches if you try and defend all the coast. So you've got to allow cliffs like this one to erode and to feed the sand into the beach. And this coast has been going back at about half a metre a year for 6,000 years and I think it's pretty arrogant of us to try and stop it now.

> **Man** – I can't see why the cost effectiveness should come into this at all. During the war we fought to save the country. Now the North Sea is taking it away.

> **Dorothy Megitt** – This isn't just the land that is going. An awful lot of the agricultural land is going. But this is England that's going.

3 Using the speech bubbles above, give three reasons why a coastline should be protected and three reasons why it should be left for nature to take its course. What are your views?

When writing case study answers, give specifics (e.g. names and dates).

Weather and climate

THE BARE BONES

➤ The weather is the daily conditions in an area. Climate is weather patterns over a long time.

➤ The angle of the sun and rotation of the earth are the main factors affecting world climate patterns.

A Weather and climate

KEY FACT

1 Weather is the condition of the lower atmosphere at one time and in one place.

- Weather conditions include **air temperature, air pressure, precipitation, wind speed, wind direction, cloud** and **sun**.

KEY FACT

2 Climate is the average or expected pattern of weather for a place, based on weather records over a long time.

A **climate graph** shows average monthly temperatures and rainfalls for a particular place.

- The data is collected over many years.

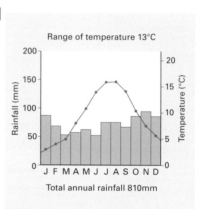

Q What is the difference between weather and climate?

B World climate zones

KEY FACT

1 The world is divided into major climatic zones. These are based on variations in precipitation and temperature between places.

Remember
Five factors affect climate.

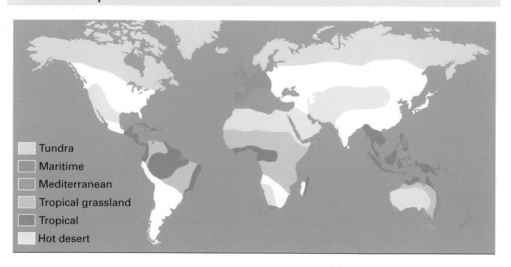

- Tundra
- Maritime
- Mediterranean
- Tropical grassland
- Tropical
- Hot desert

Q Can you name five world climate zones?

B

2 There are five main factors that affect climate:

Latitude

The angle of the sun and rotation of the earth affect the temperature and climate of different parts of the world. The sun's rays hit the earth overhead near the equator all year round, making these places hot. The sun's rays hit the poles at an oblique angle all year round, making these places cold. Whichever part of the earth is tilted towards the sun experiences summer, while the other hemisphere gets winter.

 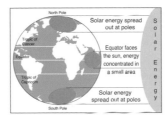 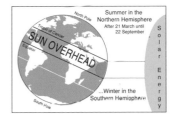

Continentality

The sea heats up more slowly and cools down more slowly than land. Places closer to the sea therefore have a smaller temperature range than those inland.

Altitude

Temperature decreases 0.6 °C every 100 metres above sea level. In addition, places with hills experience greater rainfall due to relief.

Ocean currents

Currents spread warm water from the tropics to the poles and cold water from the poles to the tropics. This affects the temperature of air passing over the currents.

Prevailing wind

This is the most common wind in an area. If the wind is tropical, it raises temperatures. If the wind is polar, it lowers temperatures.

Remember
Make sure you know which aspects of weather are in your course.

Q Can you describe a continental climate?

PRACTICE

1 Explain why it is colder at the poles than at the equator.

2 Study the map and graphs of European climate.
   a) Why do July temperatures in Berlin and Rome differ?
   b) Why is the difference between summer and winter temperatures larger at Berlin than Shannon?
   c) Which place has the highest annual precipitation?

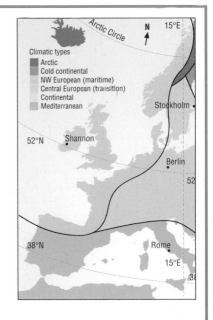

You need to be able to explain the patterns shown on a climate graph.

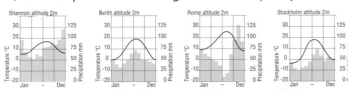

Weather and people 1

THE BARE BONES
➤ Weather and climate affect where people live and human activities.
➤ Extreme low-pressure weather conditions create hazards such as hurricanes.

A Extreme weather – hurricanes

KEY FACT

1 A <u>hurricane</u> is a low pressure system, or depression. It is sometimes called a <u>tropical storm</u> or a <u>tropical cyclone</u>.

- Hurricanes develop in the tropics above a **warm sea** (over 27 °C) and move away from the equator **westwards**.

KEY FACT

2 Hurricanes happen when the weather is <u>hottest</u>. This is between May and November in the Northern hemisphere and November and April in the Southern hemisphere.

Remember
Hurricanes are low pressure systems and form over oceans in the tropics.

3 Areas **vulnerable** to hurricanes include: the Indian subcontinent, South East Asia, Central America and the Caribbean.

4 A hurricane can stretch up to **800 km** in diameter and be up to **20 km** in height.

- The **eye of the storm** is made up of very low pressure and gives **calm weather**.

- Either side of the vortex, low pressure encourages air to rise and cool, clouds to form and rain to fall. The lower the pressure, the faster the **winds**, often over **200km/h**.

- Hurricanes **weaken** and blow themselves out when they meet **land**, as they are cut off from the warm sea that is their main source of energy.

5 The **effect on humans** can be **devastating**.

- **Winds** destroy buildings, crops, trees and power lines.

- Heavy rainfall causes river **flooding and landslides**.

- Low pressure temporarily raises the water level, creating storm surges that **flood coastal areas**.

Q Can you name three areas vulnerable to hurricanes?

A Warm moist air rises and forms cumulus clouds. Trade winds sweep in below, creating a spiral.
B Winds of over 200mph flow out of the top of the hurricane.
C Bands of rain spiral in towards the eye of the hurricane. Heaviest rain at the wall of the eye.
D The spiral forms a giant Catherine wheel with winds of 200mph near the centre.

Sequence of weather
1 Sky becomes cloudy, wind increases, sea gets choppy.
2 Cumulus clouds build up, light rain, strong winds with gusts, heavy seas.
3 Towering cumulonimbus clouds, heavy rain, strong winds and very rough seas.
4 Clear skies, warmer, no clouds or winds.
5 Towering cumulonimbus clouds again, heavy rain, strongest winds from different direction due to spiralling, very rough seas.
6 Lighter rain, wind speeds decrease, with gusts, rough seas.
7 Sky clearer, wind speeds decrease, seas calmer.

B Case study – the course of Hurricane Gilbert

1 The **National Hurricane Centre** in Miami, Florida, USA, issued a warning on 10 September 1988 that a tropical storm in the Atlantic had intensified into Hurricane Gilbert. The forecasters predicted that it was likely to affect the Caribbean and the Gulf of Mexico.

2 Hurricane Gilbert swept across this area between 9 and 14 September. It proved to be the most powerful hurricane with the lowest pressure ever recorded. **Winds** of over 200kph blew for **five days** and **waves** reached **15-metres** high.

3 It devastated St Lucia, Puerto Rico, the Dominican Republic, Haiti, Venezuela, Jamaica, Honduras, El Salvador and Mexico. It left **328 dead**, **500 000 homeless** and caused billions of dollars of **damage to crops and property**.

Remember
Hurricane Gilbert is a good example of a natural hazard.

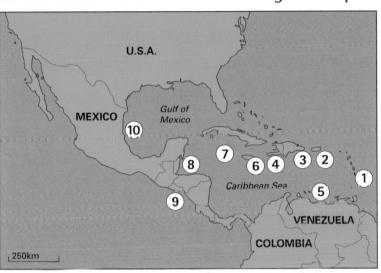

Q How many countries did the hurricane pass through?

① **9 September**: St Lucia
$2.4 million damage from rainfall

② **9/10 September**: Puerto Rico
$15 million crop damage from rainfall

③ **10 September**: Dominican Republic
$ millions of damage, flooding killed nine

④ **10 September**: Haiti
widespread flooding, 29 killed

⑤ **10 September**: Venezuela
torrential rain, floods and mudslides, five killed

⑥ **12 September**: Jamaica
500,000 homeless, 70% homes damaged, $ billions total damage, 36 killed, looters shot to death and many arrests

⑦ **13 September**: Cayman Islands
$ millions of damage, 50 homeless

⑧ **14 September**: Honduras
flooding, 6000 homeless, 13 killed

⑨ **15 September**: El Salvador
flooding, 15 000 homeless

⑩ **14 September onwards**: Mexico
26 killed during Gilbert's sweep over Yucatan Peninsula, 210 killed when northern city of Monterey devastated a few days later

Data from *The Independent* 19 September 1988

PRACTICE

1 Name a hurricane and one place where this hazard has been a problem.

2 Draw a labelled sketch map of the course of the hurricane.

3 Explain why the hazard happens there.

4 Describe how it affects people in the area.

➤ Extreme high-pressure weather conditions create hazards (e.g. droughts).

➤ Most natural droughts occur in semi-arid climates.

➤ Humans can also create drought conditions through the overuse of water.

A Extreme weather – droughts

KEY FACT

1   A <u>drought</u> is a long period of dry weather.

- Droughts develop in **semi-arid** climates in places experiencing anticyclone weather systems for long periods of time.

- These bring sub-tropical **high pressure** that holds water vapour and blows hot dry air outwards.

- The anticyclones also block depressions that bring moist air to the land.

Remember
Most droughts happen in semi-arid regions.

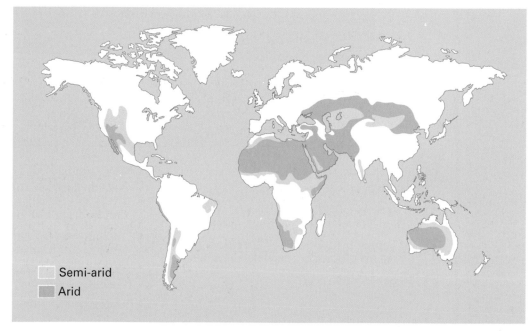

Semi-arid
Arid

2 **People** can also induce drought conditions:

- by **over-use** of existing water supplies in homes, industry and agriculture

- by **deforestation**. Cutting down trees reduces evapotranspiration, and therefore the amount of water vapour that can be released into the air. In turn, this limits clouds forming and rain falling.

3 The effects on the **environment** and people can be devastating.

- Overgrazing causes **soil erosion**.

- Crop failure, starvation of animals and water shortages cause **famine** and death.

- Increased **migration** from drought areas creates overcrowding in towns and cities.

Q What weather conditions create droughts?

B Case study – droughts in the Sahel

1 The <u>Sahel</u> region stretches from west to east across <u>Africa</u>. It is located just south of the Sahara desert and includes some of the world's <u>poorest</u> countries.

- The region includes Mali and Sudan. It is populated by **nomads**.

Remember
The Sahel experiences drought.

2 For centuries, the Sahel has fluctuated between periods of adequate rainfall and drought. The most recent serious drought began when the rains failed in **1968** and continued until the late **1980s**.

3 Continued hot, dry air from the sub-tropical high-pressure zone over the region caused widespread **land degradation**.

4 The result was mass migration south of nomads and their livestock into the savannah.

5 The increase in population, caused by the nomads' migration and high birth rates, put pressure on the fragile savannah ecosystem.

- Trees were stripped for firewood causing **soil erosion**. Now these areas are uninhabitable.

6 To avoid malnutrition and famine, the nomads move on to towns and cities as **refugees**, creating urban problems.

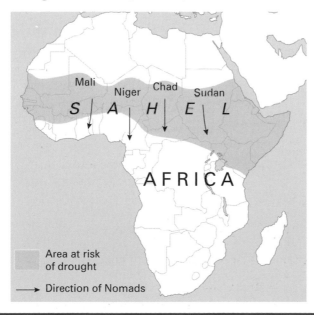

Mali Niger Chad Sudan
S A H E L
AFRICA

☐ Area at risk of drought
→ Direction of Nomads

Q Where is the Sahel?

PRACTICE

1 What is a drought?

2 How do humans contribute to drought conditions?

3 Why is the Sahel prone to drought?

4 What are the impacts of drought on the nomads?

> Make sure you learn a case study of an extreme weather event.

THE BARE BONES

➤ Human activities in industrialised countries are changing the world's climate.

➤ Global temperatures are rising, creating unpredictable weather systems and rising sea levels.

A The greenhouse effect and global warming

KEY FACT

1 The greenhouse effect is the way the <u>atmosphere traps heat</u> from the sun that is reflected back off the earth's surface.

Remember
The earth needs some greenhouse gases to function.

- Gases in the atmosphere such as **carbon dioxide, nitrous oxide** and **methane** hold in the heat.

- This keeps the surface of the earth warm. Without this greenhouse effect, the earth would be too cold for life.

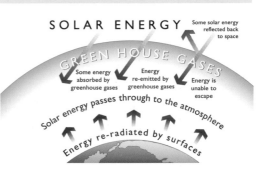

SOLAR ENERGY — Some solar energy reflected back to space

GREEN HOUSE GASES

Some energy absorbed by greenhouse gases

Energy re-emitted by greenhouse gases

Energy is unable to escape

Solar energy passes through to the atmosphere

Energy re-radiated by surfaces

KEY FACTS

2 <u>People</u> are adding more greenhouse gases to the atmosphere than occur naturally.

3 These gases trap <u>additional heat</u>, increasing the greenhouse effect so that the earth gets hotter, causing global warming.

Methane is released into the atmosphere from farming and landfill sites.

Deforestation reduces the amount of trees available to convert carbon dioxide to oxygen.

Carbon dioxide is released into the atmosphere from factories, power stations and vehicles.

Sources of Greenhouse Gases

CFC gases or chlorofluorocarbons are released into the atmosphere from aerosols, old fridges and fast-food packaging.

Nitrous gases are released into the atmosphere from fertilisers.

Q Can you draw a diagram of the greenhouse effect?

B The consequences of global warming

1 <u>Global temperatures have increased</u> by 0.6 °C in the past one hundred years.

2 Over time this will change the world's climate zones:

a) **Dry** areas such as the Sahara could spread further north to Europe.

b) **Cold** areas such as Alaska could warm up and sustain agriculture.

3 Weather patterns are changing. The USA will become drier, while Britain will become wetter in winter but much warmer in summer. **Extreme weather** events will be more common.

4 Sea levels are rising.

• The **polar ice-caps are melting**. Sea levels have risen 25 cm in the past one hundred years and are set to rise by a further 50 cm by 2100.

• Low-lying parts of the world, such as the Ganges Delta in Bangladesh and southern England, could **flood** permanently.

Make sure you can explain the greenhouse effect and show how it contributes to global warming.

Q Can you list the five main greenhouse gases?

C Slowing down global warming

Remember
Burning fossil fuels creates extra greenhouse gases.

1 This is achieved by **reducing** the number of greenhouse **gases**, particularly carbon dioxide in the atmosphere by:

• increasing **afforestation** (planting trees) and reducing deforestation

• international co-operation between countries to reduce emissions. The **1997 Kyoto Agreement** agreed to do just this (see Energy, p.128).

2 Some countries, particularly those in <u>Europe,</u> have started to <u>limit emissions</u> of greenhouse gases.

• Some LEDCs are concerned that cutting emissions would slow their rate of development and do not want to pay their own economic price for a global problem that MEDCs created.

Q How can global warming be slowed down?

PRACTICE

1 What is the evidence for global warming?

2 What causes the greenhouse effect?

3 Explain three effects of global warming.

4 Describe the future for Britain if global warming continues.

THE BARE BONES
- ➤ Weather is shown using synoptic charts and satellite images.
- ➤ Pressure is shown on maps and charts using isobars.

A Measuring the weather

1 There are six main **weather components** that need to be recorded.

Remember
Know the names of instruments that measure weather.

Temperature: Measured with a **thermometer** and recorded in **degrees Celsius**. It records the highest and lowest extremes each day.

Precipitation: Measured with a **rain gauge** recorded in **millimetres**. It records the amount of water falling from the atmosphere as rain, hail, snow, etc.

Air pressure: Measured with a **barometer**, recorded in **millibars**. It records the weight of the atmosphere pressing on the surface of the earth as high or low pressure.

Weather components

Wind: The **direction** the wind has come from is measured with a **wind vane**. Wind **speed** is measured with an **anemometer in knots or kilometres per hour**. It records the sideways movement of air from high to low pressure.

Q Can you describe how each weather component is recorded?

Relative humidity: Measured with a **wet and dry bulb thermometer**, recorded as a **percentage**. It records the amount of water vapour in the air relative to the maximum amount that could be held.

Clouds: Measured by **observation** recorded in **oktas**. It records cloud cover in the sky.

KEY FACT

2 Meteorologists plot the information collected about weather at <u>weather stations</u> on maps called <u>synoptic charts</u>.

- These are **used by forecasters** to create the maps and broadcasts that you see on television and the Internet, read in newspapers or hear on the radio.

B Reading the weather: synoptic charts and satellite images

KEY FACTS

1 Synoptic charts show weather information in <u>symbol</u> form.

2 Atmospheric pressure is shown on charts by <u>isobars</u>.

- Isobars are lines joining places of **equal pressure** at 4 or 8 millibar intervals.
- Low pressure is warm air rising. It is shown by a series of isobars in a circular shape. The **lowest value** is in the **centre**.
- **High pressure is cool air sinking**. It is also shown by a series of isobars in a circular shape. The **highest value** is in the **centre**.

Q What is a prevailing wind?

B

3 A <u>front</u> is where warm and cold air meet.

- A front is an **imaginary line** drawn by a forecaster on a weather map to separate cold and warm air.
- A **warm front** occurs when warm tropical air meets cold polar air.
- A **cold front** occurs when cold polar air meets warm tropical air.
- An **occluded front** is where a warm and cold front merge.

4 Weather <u>satellite</u> images are <u>photographs</u> taken by satellites scanning the earth <u>from space</u>.

- Black-and-white **photographs** show light surfaces that reflect light, such as clouds, as white and dark surfaces that absorb light as black.
- Colour satellite images show the temperature of the earth's surface. The darker the colour, the colder the temperature.

Make sure you can interpret synoptic charts and satellite images. General points to remember include:
- warm and cold fronts occur on the edge of the cloud patterns
- there is a vortex in the centre of low pressure with clouds circling around
- isobars follow the cloud patterns
- closely packed isobars mean low pressure and high wind speed.

Remember
Wind direction is where the wind blows from.

Q What is an isobar?

Study the satellite image below.

1 Which of the two areas marked A and B is cloudy and which is fine?

2 What type of weather system is to the west of Britain?

3 Pressure is high in area A and low in area B. From which direction is the wind blowing in the south of England?

4 Explain any similarities that you can see between the cloud pattern shown on the satellite image and the pressure and fronts shown on the weather map.

Meteosat image of Europe, 12 January 1994

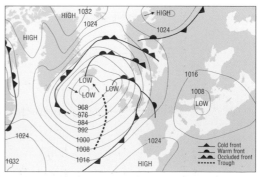

Weather map of Europe, 12 January 1994

UK weather

THE BARE BONES
➤ The weather and climate of the UK are heavily influenced by air masses.
➤ The main types of rain in the UK are frontal and relief rainfall.

A Air masses

KEY FACT ➤ **1** Air masses are very large volumes of air with <u>uniform temperature and humidity</u>.

2 Where the air **comes from** and what it **passes over** influences the weather that the air mass brings. This is because air takes on some of the properties of the surfaces that it travels over.

KEY FACT ➤ **3** The UK has changeable weather because several <u>different air masses</u> affect the country.

Remember
Air masses create most UK weather patterns.

Q Can you name five air masses?

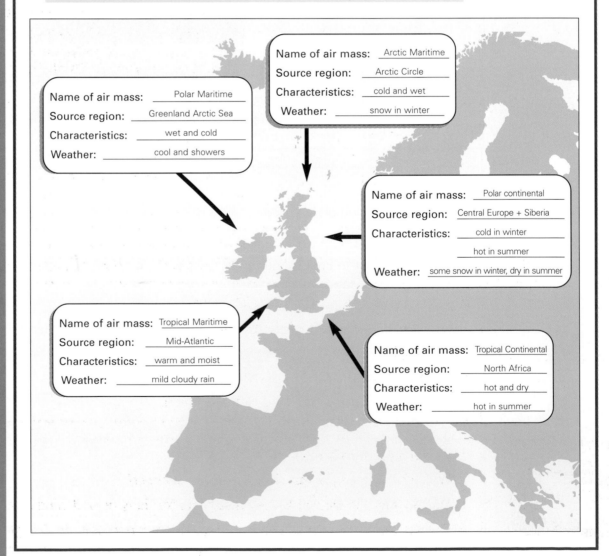

Name of air mass: _____ Arctic Maritime _____
Source region: _____ Arctic Circle _____
Characteristics: _____ cold and wet _____
Weather: _____ snow in winter _____

Name of air mass: _____ Polar Maritime _____
Source region: _____ Greenland Arctic Sea _____
Characteristics: _____ wet and cold _____
Weather: _____ cool and showers _____

Name of air mass: _____ Polar continental _____
Source region: _____ Central Europe + Siberia _____
Characteristics: _____ cold in winter _____
_____ hot in summer _____
Weather: _____ some snow in winter, dry in summer _____

Name of air mass: _____ Tropical Maritime _____
Source region: _____ Mid-Atlantic _____
Characteristics: _____ warm and moist _____
Weather: _____ mild cloudy rain _____

Name of air mass: _____ Tropical Continental _____
Source region: _____ North Africa _____
Characteristics: _____ hot and dry _____
Weather: _____ hot in summer _____

B   Types of rain

1   <u>Precipitation</u> is a term used to describe all types of <u>moisture</u> that fall from the sky.

It happens when:

- **moist air rises**
- **water vapour** in the air **cools** and **condenses** and **forms clouds**
- **droplets** grow to a certain size and then **gravity** makes them fall.

2 There are **three** main types of rainfall:

Mountain

Relief rainfall: Warm moist air cools and condenses to form clouds as it is forced to rise over **hills**. Rain falls over the **hills**. Dry air falls and warms up evaporating clouds. This creates a **rainshadow**. This type of rainfall moves west to east across the UK.

Sun's heat

Convectional rainfall: When it is **hot, air** warms and rises. As it rises, the air cools and water vapour in the air condenses to form storm clouds. These release intense precipitation, and usually **thunder and lightning**. This type of rainfall happens in the summer, often in the south of the UK.

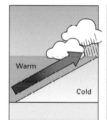
Warm   Cold

Frontal rainfall: When **two air masses meet**, warm moist air rises over colder air. This air is forced upwards. It cools and condenses, forming clouds that release rain. This type of rainfall is common all over the UK and at all times of the year.

Q How does it rain?

C   Factors influencing our weather and climate

Q Can you name the effects of land and sea on our weather?

1 Four other factors affect Britain's climate:

- Its **position** 50–58°N of the equator.
- The height of **relief**.
- Distance from the sea (**continentality**).
- **Ocean currents** (The North Atlantic Drift).

PRACTICE

1 Which air masses bring rain?

2 Which air masses are associated with the summer?

3 Explain why the west of Britain is suitable for dairying and wind farms.

4 Explain why the south and east of England are more prone to drought.

THE BARE BONES
➤ Depressions are areas of low pressure that bring unsettled, cloudy weather.
➤ Anticyclones are areas of high pressure that bring calm, clear weather.

A Depressions

KEY FACT

1 Depresssions are areas of <u>low pressure</u> that develop to the <u>west of the UK</u>.

2 Depressions are associated with **warm rising air**. Depressions, or **lows**, are created when a warm air mass and a cold air mass meet and are blown eastwards across the UK.

3 A depression forms when:

- Warm air rises up over cold air to form a **warm front**.

- This rising air results in less air at the surface, creating low pressure.

- Cold air behind the warm air forms a **cold front**.

- The cold front moves more quickly than the warm front.

- The two gradually merge forming an **occluded front**.

- Winds blow out from the centre of the low-pressure cell **anti-clockwise**. Fast winds indicate low pressure.

COLD FRONT	WARM FRONT
Warm air / Cool air	Cool air / Warm air

Earth's surface

996 992 988 984 990 984 988 992

Cool air **L O W** Cool air

Warm air

| Cool air undercuts warm air forcing it upwards at the cold front | Black arrows - warm air White arrows - cool air | Warm air rises over cool air at the warm front |

KEY FACT

4 The weather in a depression generally moves across the UK in a north-east direction and gives very <u>changeable weather</u> and <u>frontal rainfall</u>.

- As the warm front approaches, it rains.

- As the warm sector passes over, it becomes warmer and brighter.

- The occluded front brings continuous rain.

Remember
Depressions are low-pressure cells.

- As the cold front approaches, it becomes colder and windier.

- As the cold front passes over there is heavy rain.

Q Can you describe the passage of a 'low'?

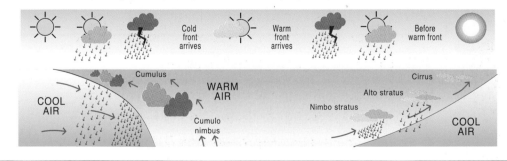

B Anticyclones

1 An anticyclone is an area of <u>high pressure</u>.

2 Anticyclones are associated with **cool**, **sinking air**. Anticyclones, or **highs**, bring settled weather that lasts for days or weeks and blocks out depressions.

Remember
Anticyclones are high-pressure cells.

3 An anticyclone forms when cold air sinks and warms. When air is warming, no clouds can form.

- Winds are light and blow out from the centre of the high-pressure cell **clockwise**.

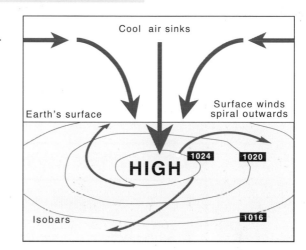

Cool air sinks

Earth's surface

Surface winds spiral outwards

HIGH 1024 1020

Isobars 1016

4 <u>Winter</u> anticyclones bring cold, clear, sunny, settled weather.

- At night there is rapid heat loss as there are few clouds. This makes temperatures very cold. This can bring **fog**, which is dense cloud at ground level, **mist**, which is thin fog, and **frost**, which is frozen water droplets on the surface.

5 <u>Summer</u> anticyclones bring hot, clear, sunny, settled weather.

- Temperatures often reach 25 °C. These high day temperatures can cause **convection rainfall** in late afternoon. Falling cool air can mix with smoke and pollution to produce **smog** over towns and cities. At night there is rapid heat loss as there are few clouds. This makes temperatures cold. This can bring morning **dew**, which is water vapour on the surface.

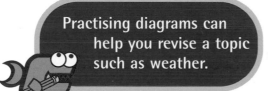

Practising diagrams can help you revise a topic such as weather.

Q Can you describe how a 'high' forms?

PRACTICE

1 Frontal rainfall is associated with which pressure type?

2 When does convection rainfall happen?

3 What cloud types are found at a warm front?

4 What cloud types are found at a cold front?

5 Describe the weather experienced during this depression.

Ecosystems

THE BARE BONES

➤ An ecosystem is an ecological system.
➤ All the parts of an ecosystem interact with each other and are interdependent.
➤ Ecosystems have inputs, processes and outputs.
➤ There are two important processes within an ecosystem: energy flows and nutrient cycles.

A What is an ecosystem?

KEY FACT

1 An ecosystem is an <u>ecological system</u>.

2 An ecosystem is a **system of plants and animals** that **live** together in a **particular environment**. The animals and plants **interact** with each other and their surrounding environment.

3 The **size** of an ecosystem can vary from very large to very small. A garden pond is an ecosystem and so is a tropical rainforest.

4 The world's largest ecosystems are called **biomes**.

Remember
You have probably studied ecosystems in science. Your science work may give some more case studies of different ecosystems.

Q Can you explain what is meant by an ecosystem?

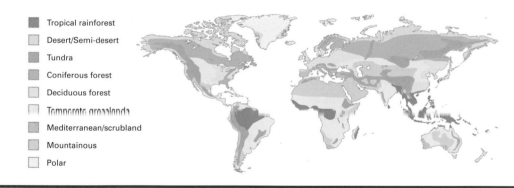

- Tropical rainforest
- Desert/Semi-desert
- Tundra
- Coniferous forest
- Deciduous forest
- Temperate grassland
- Mediterranean/scrubland
- Mountainous
- Polar

B How do ecosystems work?

KEY FACT

1 The <u>living</u> parts of an ecosystem are called <u>biotic</u>.
The <u>non-living</u> parts of an ecosystem are called <u>abiotic</u>.

2 The living and non-living parts of an ecosystem interact with each other. They are interdependent, which means that they depend on each other to make the system work.

KEY FACT

3 If one part of the ecosystem is altered, the whole system will be affected.

• The abiotic (non-living) part of the ecosystem provides the environment that helps plants and animals of the ecosystem to survive.

B

4 Like many other geographical systems, an ecosystem has inputs, processes and outputs (see the river system, p.28).

- The **greater** the **inputs** into the ecosystem, the **greater** the **diversity** of plants and animals **supported** by the ecosystem. For example, a desert has limited inputs of rain and soil nutrients and so the range of vegetation and wildlife found in a desert is small.

5 There are two important **processes** within an **ecosystem**: **nutrient cycles** and **energy flows**.

- Sometimes the flows within an ecosystem are shown using arrows. The arrows show the direction of the flow and how great the flow is. The wider or larger the arrow, the greater the flow of nutrients of energy.

Nutrient cycles

- Nutrients are found **naturally** in the rocks, water and the atmosphere.
- Nutrients are important **minerals** such as nitrogen, magnesium, calcium, phosphorous and potassium. These nutrients are essential to all living things.
- The **nutrient cycle** moves these minerals through the ecosystem.
- **Decomposers**, such as small bacteria and fungi, break up dead leaves and organisms and return the nutrients to the soil, to be taken up by plants and trees.
- Nutrients may be **lost** from the ecosystem. This may happen if the **nutrients** are **washed** away by **surface runoff** or through **leaching**. Leaching occurs when the nutrients are **washed down** to the lower **layers** of the **soil**, where they can't be reached by plants or trees.

Energy flows

- Energy flows through an ecosystem through the **food chain**.

All energy comes from the sun. Plants are **producers**, making plant food (glucose) through photosynthesis.	→	Herbivores eat plants. These are **primary consumers**.	→	Herbivores are eaten by carnivores. These are **secondary consumers**.	→	Energy is lost when plants and animals breathe. More energy is lost higher up the chain.

Remember
Within an ecosystem there are other smaller systems at work, for example the nutrient cycle.

Q Can you explain what leaching means?

PRACTICE

1 Use the map opposite to name three types of ecosystem.

2 Briefly explain why there are few plants or animals found in a desert ecosystem.

3 With reference to a named ecosystem, explain how energy flows through the system.

Make sure you can draw a food chain for an ecosystem that you have studied

THE BARE BONES

➤ Rainforest ecosystems are found close to or along the equator.

➤ Rainforests are found in areas of tropical climate. This provides the system with high inputs of rain and sun.

➤ Rainforest ecosystems can support a rich diversity of plant and animal species.

A The location of tropical rainforests

KEY FACT

Tropical rainforests are found in a <u>belt 5° north or south</u> of the equator.

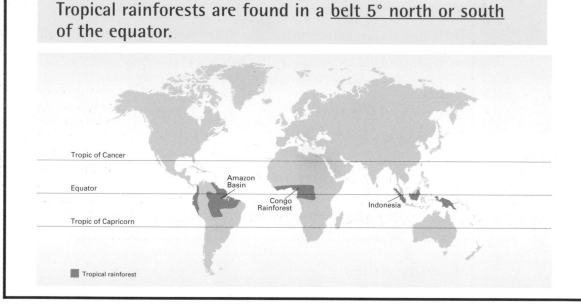

Tropic of Cancer

Equator

Amazon Basin

Congo Rainforest

Indonesia

Tropic of Capricorn

■ Tropical rainforest

Q Use the map opposite and an atlas to help you name some of the world's rainforests.

B How does the rainforest ecosystem work?

1 The average **daily temperature** in the rainforest is between **25–30 °C**, while the average **yearly rainfall** is between **1500mm** and **3000mm**.

2 This means that the rainforest ecosystem has very **high inputs of rain and sun**. The atmosphere is very **moist and humid**.

3 These conditions are excellent for <u>plant growth</u>, which means that rainforests have a <u>high diversity of plants, insects, birds and animals</u>.

KEY FACT

Remember
If you are describing the location of the world's rainforests, name specific places, such as the Amazon Rainforest (Brazil).

4 **Nutrient cycling** in the rainforest is very **rapid**, because the **humid conditions** help dead matter to **decompose** quickly. There is also a vast network of **bacteria** and **fungi** that helps break down dead material and return **nutrients** to the **soil**. These nutrients help to support the plant and animal species in the forest.

5 The **structure of the rainforest** is very important in helping the **ecosystem** to **function**. Rainforests have distinct layers of vegetation.

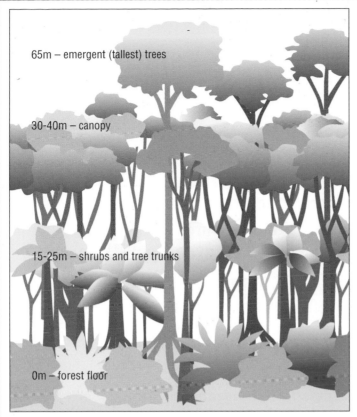

65m – emergent (tallest) trees

30-40m – canopy

15-25m – shrubs and tree trunks

0m – forest floor

The emergents: These are the tallest trees that have emerged out of the canopy layer. These trees can grow up to 65 metres tall.

The canopy: This layer forms the 'roof' of the forest. The trees are up to 40 metres tall and have grown high upwards in search of light. Small climbing plants, such as lianas, are often wrapped around the trunks of these trees.

The lower canopy: Here are the youngest trees and saplings. It is still quite dark here and the trees are fighting for the light.

The forest floor: Here there is less vegetation due to the dark, damp conditions. This layer has a thick layer of decomposing leaves and is home to the large buttress roots of the largest trees.

Q Can you describe the inputs into the rainforest system?

C Why are rainforests being destroyed?

1 The rainforest <u>ecosystems</u> are <u>under threat</u> from destruction by <u>humans</u>. Many countries and businesses want to take advantage of and <u>exploit</u> the resources of the forest.

2 The Amazon rainforest is the world's largest rainforest and areas of it are being destroyed as people make use of its resources.

Logging: Expensive wood, such as mahogany, is cut down by logging companies and exported across the world.

Cattle ranching: Large areas of the forest have been cleared for raising cattle, which can then be sold for their meat.

Reasons for the destruction of the Amazon rainforest

Roads and buildings: Land in the forest needs to be cleared to make space for industrial buildings, accommodation for the workers and for building roads to get goods in and out of the region.

Mining: In the Amazon there are huge reserves of minerals such as gold, iron ore and copper. These are used in industries throughout the world. To get at these minerals, vast stretches of the forest must be destroyed and mines excavated.

Remember
Rainforest ecosystems rely on high inputs. These inputs help create a very rich and diverse ecosystem.

Q Can you list three reasons for the destruction of areas of tropical rainforest?

THE BARE BONES

➤ The diversity of the rainforest ecosystem means that it provides humans with many resources. This has led to the destruction of many rainforest areas.

A What are the effects of deforestation in the rainforest?

KEY FACTS

1 <u>Deforestation</u> within a rainforest can have many impacts on both <u>people</u> and the <u>environment</u>.

2 <u>Cutting down trees</u> in the rainforest can <u>destroy the ecosystem</u>. If one part of the system is removed, it can no longer work or function properly.

• Here are some of the key effects of rainforest deforestation:

Removing the trees from the rainforest system affects the **natural processes** taking place within the ecosystem.

People think that because the **vegetation** in the rainforest is **lush,** then the **soils** must be very **fertile. This is not true.** The growth of rainforest vegetation takes place because of the **nutrient cycle.**

3 The **nutrient cycle** moves these minerals through the ecosystem:

Remember
Not all of the world's rainforests are being destroyed. In places where the forests are being destroyed, people cannot agree on how fast this is happening.

Decomposers (insects, bacteria and fungi) **break down dead matter** (animals and plants) and **return the nutrients to the soil.**

Plants take up **nutrients** from the soil and use them to grow.

Animals eat the plants.

Nutrients are **stored** within the ecosystem by plants and animals (**biomass**). Nutrients are also stored in the **leaf litter** (leaves found on the floor of the ecosystem).

• If trees are removed, the cycle is **broken**. The soil becomes infertile and nutrients stored in the soil are washed away by the rain. This is called **leaching**. Removal of trees leaves the **soil exposed**, causing erosion by wind or rain.

4 Removing trees from a tropical rainforest affects the amount and type of **plant and animal species** in the ecosystem. Deforestation destroys animal habitats and reduces the flow of energy within the system. Over time, plant and animal species may reduce in number and eventually become **extinct**. Some species of rainforest vegetation provide humans with many common household **medicines**. Some experts believe that certain rainforest plants could provide cures for disease.

Q Can you list three impacts of deforestation in the rainforest ecosystem?

5 Many scientists believe that deforestation of rainforests could lead to changes in the world's **climate**. Trees absorb carbon dioxide and give out oxygen. Cutting down trees alters the balance of gases in the atmosphere. Increased greenhouse gases such as **carbon dioxide** adds to the problem of **global warming**.

B How can the rainforests be managed?

1 Scientists cannot agree on <u>how fast</u> the world's rainforests are being destroyed.

2 Some people **predict** that if we continue to remove trees from the world's rainforest ecosystems, the forests will have **disappeared** within the next hundred years. However, some experts **disagree** with this view and argue that there are many large areas of **undisturbed** rainforest across the **world**.

3 Although people can't agree on how fast the world's rainforests are being destroyed, more and more governments and international organisations have begun to **realise their importance**.

4 Rainforests supply the world with a valuable supply of **oxygen** and contain important plant and animal species. Therefore, measures have been taken to protect the world's rainforest ecosystems:

5 **Sustainable logging** is taking place in many areas. Once a tree is cut down, another is planted.

6 Areas of rainforest have been **designated as protected areas**, where no development can take place.

7 People have been **educated** about the importance of the rainforest and **consumers** in MEDCs are **given information** about where the wood they buy has come from.

8 In parts of the Amazon, **mining companies** are required to **replant trees** once they have finished mining the area.

> If you are asked to write about a rainforest ecosystem, make sure you name the rainforest you are writing about.

Q Can you list four ways in which people can manage a rainforest ecosystem like the Amazon?

PRACTICE

1 Use the map on p.68 to describe the distribution of the world's rainforests.

2 Give two reasons for the rich diversity of wildlife in the rainforest ecosystem.

3 For a named ecosystem that you have studied, explain why humans have changed the ecosystem and describe what impacts this has had on the environment.

Population distribution and density

THE BARE BONES
- Patterns of population distribution tell us where people live.
- Some areas of the world are densely populated, while others are sparsely populated.
- Some natural environments attract large numbers of settlers, while other environments discourage settlers.

A Population distribution

KEY FACTS

1 Population means the total number of people living in an area.

2 Population distribution means the way in which people are spread across a particular area. The scale of this area can be local, regional or global.

3 Patterns of population distribution tend to be uneven. Some areas are densely populated (crowded), while others are sparsely populated (few people).

4 One way of **measuring** population distribution is to calculate the **population density** of an area. Population density is the **average** number of people living in a **square kilometre**. You can calculate population density using the following **formula**:

Population Density = Total Population ÷ Total Land Area (km^2)

5 **Patterns** of population **distribution** and **density** are often shown using a **choropleth** map. A choropleth map **uses shading** to show different values. For example, on a population density map, the darker the shading, the higher the population density.

World population

Tropic of Cancer

Equator

Tropic of Capricorn

■ Densley populated ■ Moderately populated □ Sparsely populated

The map above shows world population distribution. The pattern of distribution is uneven, with areas of high population density (e.g. the UK and Europe) and areas of low population density (e.g. Greenland). Areas of high population density tend to be located in the northern hemisphere.

Remember
Countries with a high population density can be either MEDCs or LEDCs. LEDCs often have low population densities as many people live in small rural communities.

Q Do you know how to calculate population density?

B Factors affecting population density

1 We can **explain** patterns of **population distribution** by looking at the **natural** and **human environment** of different places.

2 The world is made up of many <u>different types of environments</u>. Some of these environments <u>attract settlers</u>, while others <u>repel</u> (discourage) settlers.

Factors attracting settlers

Temperate climate: Areas with moderate climates – plenty of rain and sunshine – (e.g. North America).

Low-lying areas: Flat, fertile land is good for agriculture and industry. River deltas and flood plains are very fertile (e.g. the Ganges Delta, Bangladesh).

Supplies of natural resources: Plentiful supplies of natural resources (e.g. coa) attract settlers. These resources provide fuel for homes and industry.

Factors repelling settlers

Extreme climates: Desert areas or extremely cold areas are hard to cultivate (farm).

Mountainous areas: Farming and building is difficult in highland areas due to poor soils, cold temperatures and a thin atmosphere (e.g. the Himalayas).

Vegetation: Dense rainforest is not easily accessible. The tropical climate is hot and very wet.

Q Can you name a sparsely populated country and a densely populated one?

C Population distribution in the UK

1 Population density is an **average** figure and it can hide differences in population density **within a country or region.** The UK has a relatively high population of 243 people per square kilometre. However, **within** the UK there are areas of high population density and areas of low population density.

2 The **north-west Highlands** of Scotland have a low population density, while **south-east England** has a high population density.

Q Describe the population distribution in the UK?

Highlands – difficult to farm and build on. Communications are difficult.

Less than 10 people per square km.

Glasgow • • Edinburgh

■ Major urban settlements

Liverpool • • Manchester

Birmingham & the West Midlands

London & the South East

Lowland areas – good climate. Ideal for farming and industry.

More than 150 people per square km.

Make sure that in your exam you read command words carefully. <u>Describe</u> means say what you see. <u>Explain</u> means say why it is like that. If you are asked to describe a pattern, you are not being asked to give any reasons for the pattern – just describe it.

1 With reference to named areas and places, describe the pattern of population distribution on the world map opposite.

2 Explain the distribution of population in the UK.

Population change

THE BARE BONES
➤ The world's population is rising rapidly. It reached 6 billion in 1999.
➤ Population growth is caused by differences between births and deaths.
➤ Population growth is highest in LEDCs.

A Population growth

KEY FACT

1 The world's population is <u>growing rapidly</u>.

2 For thousands of years, the world's population grew at a **steady rate**.

- In **1820**, the world's population reached **one billion**.

- In **1999**, less than two hundred years later, the world's population totalled **six billion**.

- Since the 1960s, the world's population has been growing at a rate of **one billion every 15 years** or so.

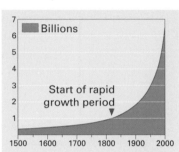

3 This **rapid rise in the world's population** has been called a **'population explosion'**.

Remember
The world's population is growing rapidly because birth rates are still very high in LEDCs, while death rates are falling across the world.

4 There are **three** causes of population change:

Births – measured using the **birth rate** (number of live births per 1000 of the population).

Deaths – measured using the **death rate** (number of deaths per 1000 of the population).

Migration – the **movement of people in and out of a country**.

Q What are the three main causes of population change?

5 The **difference** between the **birth rate** and the **death rate** of a country is called the rate of **natural population increase**.

- If the birth rate is higher than the death rate, the total population will increase.

- If the death rate is higher than the birth rate, the total population will decrease.

KEY FACT

6 Population growth rates are <u>highest in LEDCs</u>, where birth rates are high but death rates are beginning to fall. Death rates are falling due to global improvements in health and medicine.

B The demographic transition model

1 The demographic transition model opposite shows **population change over time**. It is divided into **four stages**.

2 As a country passes through the four stages of the model, the total population increases. Most **MEDCs** are in **stage four** of the model; most **LEDCs** are currently at **stage two** or **three**.

B

Q Can you explain what the demographic transition model is?

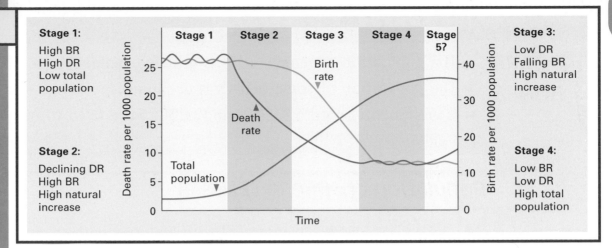

Stage 1:
High BR
High DR
Low total population

Stage 2:
Declining DR
High BR
High natural increase

Stage 3:
Low DR
Falling BR
High natural increase

Stage 4:
Low BR
Low DR
High total population

C *Population structure*

1 Population structure shows how the population of an area is divided up between males and females of different age groups.

2 **Population pyramids** are used to show population structure.

3 Population pyramids can be drawn for a whole country or individual settlements (towns or villages).

4 The **shape** of a population pyramid gives us information about birth rates, death rates and life expectancy in a country or settlement. **Life expectancy** is how long, on average, a person can expect to live.

5 A population pyramid also tells us about the number of dependants living in an area. **Young dependants** are **children under** the age of **15** who are still **supported** by their family. **Elderly dependants** are those people aged **over 65** who no longer work. Both groups depend on the economically active (those of working age).

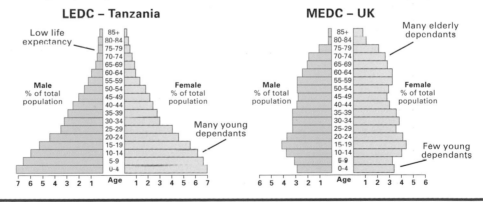

KEY FACT

Remember
Population pyramids can change over time. Two hundred years ago, the UK had a population pyramid with a wide base and a narrow tip, just like the population structure of many LECDs today.

Q Can you list four differences between the two pyramids opposite?

PRACTICE

1 Use the graph opposite to work out:

a) the population of the world in 1950.

b) what had happened to the world's population by 1999.

2 Describe the main changes that take place as a country moves through the demographic transition model.

Population issues

THE BARE BONES
- The population of many LEDCs is growing rapidly.
- The population of some MEDCs is declining.
- Governments in both MEDCs and LEDCs have tried to manage population growth.

A Population change in LEDCs

KEY FACT

1 The population of most LEDCs is <u>rising rapidly</u>.

2 Most LEDCs are now at **stage two** or **three** in the **demographic transition model** (see p.75). They have **high birth rates** and **falling death rates**. This means that LEDCs have a **high rate** of **natural increase**.

3 Death rates are falling due to modern improvements in health care and medicine.

4 There are several reasons for high birth rates in LEDCs:

Limited education about family planning.

Lack of contraception.

It may be traditional to have large families.

Causes of high birth rates

High rates of infant deaths (infant mortality) results in women having more children to ensure that some survive through to adulthood.

Children may be needed to earn money or farm the land.

It may not be culturally or religiously acceptable to use contraception.

Q Can you make a list of reasons for high birth rates in LEDCs?

5 The high rate of population growth in **LEDCs** means there are **many young dependants**. This creates problems for LEDCs. As they grow older, these dependants will need housing, healthcare and employment.

KEY FACT

6 The <u>governments</u> and <u>aid charities</u> in LEDCs want to <u>reduce birth rates</u> and slow down the rate of population growth.

B Managing population growth in Tanzania

1 Tanzania is one of the **poorest** countries in the world. It has a total population of 36.2 million.

2 The **birth rate** in Tanzania is **higher** than the death rate, so it has a **high rate of natural increase**.

3 Tanzania has a birth rate of 41, while the death rate is 13. The rate of natural increase is 2.8%.

4 If the population of Tanzania **continues** to grow at the current rate of natural increase, it will reach 57 million by the year 2025.

B Case study – Tanzania

Q Can you explain how population growth has been managed in Derada, Tanzania?

Derada is a village in Tanzania where many women have had large families, with lots of young children dying before their fifth birthday. The use of contraception is not traditionally acceptable.

The local authorities and charities working in the area have tried to reduce the birth rate in Derada. A local health clinic has been built where children are vaccinated against childhood diseases, and women are offered advice about contraception and family planning.

These changes have meant that rates of infant mortality (infant deaths) have fallen and women are having smaller families. Women are also able to plan when they have children and complete their education.

C Population change in MEDCs

1 The population **growth rate** of most MEDCs is **stable**. MEDCs tend to have **low birth rates** and **low death rates**.

2 The population of some MEDCs is actually **declining**. For example, in **Germany** the birth rate is lower than the death rate and the **rate of natural increase** is −0.1%.

3 In some MEDCs, the government is trying to **increase the population growth rate**. They have done this by providing **incentives** for young people to consider starting a family, such as **paid maternity leave** and **child benefit**.

KEY FACT

4 One of the biggest issues facing MEDCs is the problem of an <u>aging population</u>.

5 As **modern health care** becomes more and more advanced and the **standard of living** in MEDCs rises, **people are living longer**.

Case study – UK

Q List the causes of an aging population in MEDCs?

In the UK, the average life expectancy is high and, in the future, the number of people living longer than 80 years old is set to rise dramatically. This means that MEDCs such as the UK will have a high number of elderly dependants. Elderly dependants need to be provided with health care, public transport services, suitable housing and entertainment facilities. Raising enough money to care for elderly dependants could mean that taxes have to rise and a greater financial strain is placed on the economically active sector of the population.

PRACTICE

1 Why are many LEDCs experiencing rapid population growth?

2 With reference to a named example, explain how population growth can be controlled.

3 What are the impacts of an increasingly aging population?

In your exam, make sure you read questions carefully and check if you are being asked to write about an LEDC or an MEDC.

Migration

THE BARE BONES

➤ Migration is the movement from one place to go to live in another.
➤ Migration can be explained by 'push' and 'pull' factors.
➤ Some people are forced to leave their home due to war or natural disasters. These people are called refugees.

A Types of migration

Remember
To remember this, think of the letter 'I' for people going 'into' a country and the letter 'E' for people 'exiting' a country.

Q Explain the difference between internal migration and international migration.

1 There are several **different types** of **migration**:

Internal migration is when people move within a region or country.

International migration is when people move from one country to another.

Voluntary migration is when people choose to move, perhaps for a better lifestyle or job.

Forced migration is when people are forced to move home, perhaps due to a natural hazard or war.

2 People who migrate are called migrants.
• **Emigration** is when someone **leaves** a country.
• **Immigration** is when someone **moves into** a country.
• Migration can be **permanent** or **temporary**.

3 Many countries now have **laws** and **policies** to **control** the number of **people immigrating** into the country.

B Push and pull factors

KEY FACT

1 Migration is usually caused by a combination of push and pull factors.
• **Push factors** are the reasons why someone wants or needs to **leave** an area.
• **Pull factors** are the reasons that **attract** someone to a new area.

Push Factors: unemployment, crop failure, lack of services and amenities, drought, flooding, war, poverty.

Pull Factors: employment opportunities, better housing, better services and amenities, better education and healthcare.

Q List three push factors and three pull factors?

2 When making a decision, people may also consider **reasons to stay** where they are or **negative points** about the place they are planning to move to.

3 When a **large number** of **people** leave an area, it is called **depopulation**.

C Case Study — International migration from Turkey to Germany

During the 1950s and 60s, thousands of workers migrated from Turkey to West Germany. They left Turkey because of unemployment and low wages (push factors). Germany encouraged the migrants to come to Germany as 'guest workers'. Germany wanted to rebuild its economy and repair the damage done by World War II. Turkish migrants were offered work and a regular wage (pull factors).

At first, male migrants left Turkey and found work in Germany. Once they had settled in the new country, their families joined them.

Tensions began to grow between German citizens and Turkish migrants. Many Turkish migrants couldn't speak German and they sent home a lot of the money they earned to Turkey. Many Turkish workers were exploited and had to work long hours in poor conditions.

During the 1970s, Germany suffered a recession and many Turkish migrants lost their jobs. This increased tension between German people and the Turkish migrants, as the German government had to look after the unemployed Turks. The government needed to provide them with housing and other social benefits.

Since the early 1990s, the German government has tightened its controls on immigration and sometimes encourages migrants to return home to Turkey.

Remember
Migration can be permanent or temporary. Going away to study at university can be seen as temporary migration.

Q Can you describe the problems faced by people who migrated from Turkey to Germany?

D Refugees

1 <u>Refugees</u> are people who have been <u>forced</u> to leave their home.

2 Refugees may be forced to leave due to **war, political or religious conflict or natural hazards** such as floods.

3 Most refugees move to a **nearby region** or country. For example, due to conflict in Afghanistan, many people have fled to neighbouring Pakistan. Some move **further away,** to countries where they feel they will be offered political tolerance and the chance of a better quality of life. For example, during the **Kosovan conflict**, many refugees applied for asylum in the **UK** and **Germany**.

4 During the 1990s, there were estimated to be **fifty million refugees in the world**. Eighty per cent of this refugee population lives in countries outside of Europe and North America.

Q What is a refugee?

PRACTICE

1 For a named example you have studied, explain why people have migrated to another country and describe the impact of migration on that place.

Settlement

➤ A settlement is a place where people live. The size and shape can vary.

➤ The site chosen is dependent on a number of locational factors.

➤ Settlements can be ranked in a hierarchy according to size and number.

A Settlement site and situation

KEY FACT

1 A settlement can be permanent or temporary. The reason why a settlement is built is known as its function (e.g. a port or market town).

2 The **site** of a settlement will have been **chosen by settlers** because it has **one or more attractive features**:

Remember
It is essential that you know and understand the difference between site and situation.

Flat, fertile land.

Dry land (away from marshy areas).

A good water supply (e.g. a river).

Supplies of natural resources, such as coal or wood for fuel.

Attractive site features

Close to high land for defence, shelter or flood protection.

Access to building materials, such as wood or stone.

Good access and communications (e.g. a bridging point along a river).

3 The **piece of land** on which a settlement is built is known as its **site**.

Q Can you explain what is meant by the terms 'site' and 'situation'?

• The **situation** of a settlement is its **position** in relation to the **human and physical features surrounding it** (e.g. roads, railways or other settlements).

• When writing about the site of a settlement, refer to the **relief** of the land, the **vegetation** and any important **physical features** inside the settlement (e.g. a river).

B Settlement patterns

Settlements are usually arranged in a pattern. There are three basic types of pattern:

1 Nucleated: The buildings are clustered around a central point in the settlement (e.g. a bridge).

2 Dispersed: The buildings are spread out. Common in sparsely populated areas (e.g. pastoral farming regions).

Q Can you list three types of settlement pattern?

3 Linear: The buildings are arranged in a line. Often along a road, river or at the foot of a hill. Common in valleys.

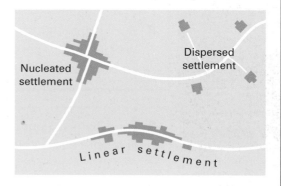

Nucleated settlement

Dispersed settlement

Linear settlement

C Settlement hierarchies

1 Settlements can be ranked in order of their size and number. This is called a <u>settlement hierarchy</u>.

2 As you move up the hierarchy, the <u>size</u> of the settlements <u>increases</u> and the <u>frequency</u> (number) <u>decreases</u>.

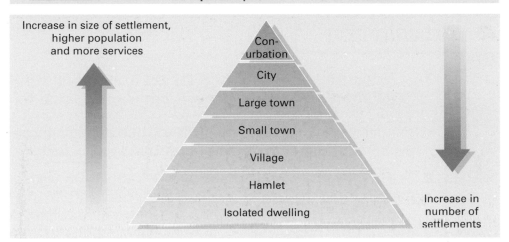

Increase in size of settlement, higher population and more services

Con-urbation
City
Large town
Small town
Village
Hamlet
Isolated dwelling

Increase in number of settlements

3 All regions or counties have a settlement hierarchy. At the bottom of the hierarchy are the smallest settlements (single settlements and hamlets). At the top of the hierarchy are larger settlements (cities or conurbations). There are **many small settlements** and **fewer large settlements**.

• A **conurbation** is a very **large urban area**, where several small towns or cities have **merged** together.

4 The higher up the hierarchy you go, the greater the range and number of **services** you will find in a settlement.

• **Hamlets** often have **no services**, except for maybe a phone box or post box.

• A **village** has a limited range of essential services, such as a post office, supermarket and doctor's. These are **low-order services**.

• **Large towns** and **cities** have a wider range of services, such as leisure centres, chain stores and universities. These are **high-order services**.

5 **Low-order services** provide **everyday goods**; high-order services tend to be used infrequently. People usually have to **travel further** to use **high-order services**.

In your exam, if you are given a question involving a map extract, always quote grid references in your answer and use the key to help you read the map.

Q. Can you explain what a settlement hierarchy is?

1 Use the OS map on p.11 to list two attractive features of the site of the settlement of Newport.

2 What is the pattern or shape of Broadstreet Common, 3584 (see p.11)?

3 What evidence is there on the map that Broadstreet is a hamlet?

Urbanisation patterns

THE BARE BONES

➤ Urbanisation means an increase in the proportion of people living in urban areas.

➤ On the global scale levels of urbanisation are increasing.

➤ The fastest rates of urbanisation are currently occurring in LEDCs.

A Urbanisation

KEY FACT

1 Urbanisation means an increase in the proportion of people living in urban areas (towns/cities) compared to rural areas.

• You can measure the **level of urbanisation** in a country or region by looking at the **percentage** of the **total population** living in **urban areas**.

Remember
A country is considered urbanised when over fifty per cent of its population live in an urban area.

2 As a country **industrialises**, levels of **urbanisation** tend to **increase**, as **people move to towns** and cities looking for work. Most **MEDCs** industrialised over a hundred years ago, so they already have **large urban populations**. Many **LEDCs** are in the **early stages of industrialisation** and so they have a **growing urban population**.

• MEDCs tend to have **high** levels of urbanisation. Ninety per cent of people in the UK live in urban areas.

• LEDCs tend to have **low** levels of urbanisation. Twenty-eight per cent of people in India live in urban areas. However, most LEDCs are **urbanising** at a **rapid rate**.

Q Can you explain what is meant by urbanisation?

3 **Urbanisation** is taking place on a **global scale**. In 1900, only about 10 per cent of the world's population lived in urban areas. Today 47 per cent of the world's population live in urban areas. If **levels** of **urbanisation** continue to **rise rapidly** in LEDCs, the **world's urban population** will **increase**.

B Millionaire cities

1 A **millionaire city** is a city with a **population** of **over one million** people.

• There are currently more than 280 millionaire cities in the world.

KEY FACT

2 The number of millionaire cities in the world is growing at a rapid rate. Most of this growth is occurring in LEDCs.

• **Most** of the **world's largest cities** are found in **LEDCs**.

3 Some cities have **over ten million inhabitants** and are called **mega-cities**, such as Mexico City, Los Angeles and Tokyo (see map opposite).

4 **Urban areas** tend to **grow outwards** and **spread out** into rural areas (countryside). This is known as **urban sprawl**.

B

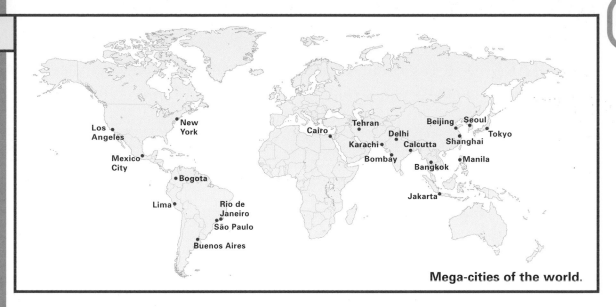

Mega-cities of the world.

Q Can you define the terms 'millionnaire city' and 'mega-city'?

C Rural-urban migration in LEDCs

1 There are **two** main causes of rapid urbanisation in LEDCs:

- **Rural to urban migration** involves the movement of people from the countryside to cities.

- There is a high rate of **natural increase in urban areas**. This is due to a **high birth rate** (migrants to the city tend to be of child-bearing age) and a **falling death rate** (healthcare facilities are better in urban areas).

2 **Push and pull factors** can be used to help explain why people in LEDCs are **moving** from the **countryside** to **town and cities**.

Push factors (encouraging people to **leave rural areas**)	Pull factors (**attracting** people to **urban areas**)
Unemployment Low wages Unprofitable farming Few employment opportunities The need to support a growing population (population pressure) Lack of social amenities and leisure	Greater number and variety of employment opportunities Higher wages Chances to improve their standard of living Better schools and hospitals Better housing and basic services (water, electricity, sewerage) Opportunities for a better social and cultural life Better transport and communications

Remember
Learn the meaning of the term 'urbanisation'. Migration to cities is not urbanisation. Migration can cause urbanisation, but don't get the two muddled up.

Q Push and pull factors: can you list four of each?

PRACTICE

1 If a country is said to be urbanising, what does this mean?

2 Use the map above to describe the distribution of mega-cities.

3 Explain why rural-to-urban migration is a common process in many LEDCs.

Make sure you understand why LEDCs are urbanising quickly.

THE BARE BONES
> It is possible to identify different zones of land use within a city.
> These zones tend to be arranged to form a land-use pattern.

A Urban land use

1 As cities **grow** and **develop**, it is possible to identify **patterns** of **land use** within them. Each of the different land uses taking place within a city tends to be found in a particular area or **zone**.

2 There are **four** land-use zones common to most cities:

 Commercial, business and administrative land uses (e.g. shops, offices and banks).

 Industrial land uses (e.g. factories and small production centres).

 Residential land use (e.g. housing).

 Open land (e.g. parkland and sports grounds).

- These categories can be **broken down** further into **smaller zones** (e.g. in a residential area there will be zones of high-class housing and low-class housing).

Q Can you name the four main urban land uses found within a city?

KEY FACT

3 The way in which these land uses are <u>arranged</u> within a town or city tends to <u>differ</u> between <u>MEDCs</u> and <u>LEDCs</u>.

B Urban land-use patterns in MEDC cities

1 To help understand patterns of land use within towns and cities, geographers have drawn **models** of a 'typical' urban settlement. These models are simplified versions of what cities are really like and they make many **generalisations**.

2 The **first models** to be drawn up were based on MEDC cities in Europe. However, they do have **limitations**. Many were developed before the growth of mass **car ownership**. Cars have allowed people more freedom to live, work and shop out of the city. This has had an impact on land use within cities.

Remember
All cities are different and land-use models can only give us a general idea about a particular city.

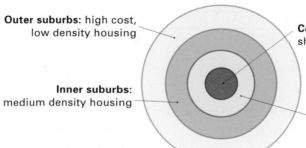

Outer suburbs: high cost, low density housing

Central Business District (CBD): shops and offices

Inner suburbs: medium density housing

Zone in transition or twilight zone: inner city - industry and high density, low cost housing

B

Q An alternative urban land-use model is the Hoyt Model. Find out about this model.

3 One of the most famous urban land-use models is the **Burgess Model**. It is based on a **pattern of concentric rings**. Each ring forms a land-use zone.

- The Burgess Model suggests that cities in **MEDCs** have grown outwards, with the **newest** and **most expensive housing** being on the **edge of the city**.
- The Burgess Model also suggests that **land values fall as you move from the centre** of the city. This means that on the edge of the city there is less competition for land and more space for development. Meanwhile, in the **CBD**, the land values are **high** and there is a lot of competition for land. This theory of declining land values with distance from the CBD is called **distance decay**.

C *Land-use patterns in LEDC cities*

1 The land in most LEDC cities is used in a similar way to MEDC cities. However, the **pattern** of these land uses is **different**.

2 In LEDCs, areas of **high-class housing** and middle-class housing tend to be found **just outside the CBD**. The poorer quality housing is found on the edge of the city.

- This pattern is the opposite to **MEDC cities**.

KEY FACT

3 The areas of poorer quality housing found on the edge of the city are called <u>squatter settlements</u> or <u>shanty towns</u>.

4 Not all LEDC cities are exactly the same, but it is still possible to draw a general **model** to show how the land is used in many of the cities.

- The model below shows patterns of land use in a typical South American LEDC city.

Favelas: recent informal housing (poor quality, may be self-built)

Periferia: older informal housing (improved over time, more permanent)

High cost housing: luxury flats or detached houses

Central Business District (CBD)

Industry: along transport routes

Q Do you know the key differences between land-use patterns in MEDCs and LEDCs?

PRACTICE

1 Explain what an urban land-use model is.

2 Explain what is meant by 'distance decay'.

3 Describe the urban land-use pattern within a typical MEDC city.

4 Describe the urban land-use pattern within a typical LEDC city.

If you are describing urban land-use patterns, relate your answer to a specific town or city that you have studied, and sketch the appropriate model.

THE BARE BONES

➤ Rapid urbanisation is taking place in many LEDCs.
➤ There is a big gap between the rich and the poor in LEDC cities.
➤ Most people in an LEDC city live in vast squatter settlements on the edge of the city. In many cities, improvements are being made to these areas.

A The quality of life in an LEDC city

KEY FACT

1 Cities are growing rapidly in many LEDCs due to <u>rural–urban migration</u> and <u>natural population growth</u>.

Q Give two reasons for urban growth in LEDCs.

2 In most LEDC cities there is a **big gap between the rich and the poor.** The rich often live in modern, high-security apartments. The poor live in **basic housing**, often on the **edge** of the city and surrounded by conditions of **poverty. The poor are often the majority** in an LEDC city.

B Squatter settlements

1 Every day, **migrants** arrive in the city looking for somewhere to live. They tend to **settle** on the **edge** of the city in **squatter settlements**.

2 Often countries have their own names for squatter settlements. For example, in Brazil they are known as *favelas*. In India they are called *bustees*.

3 Squatter settlements are usually **illegal** and found on **poor-quality land**. They tend to be **unplanned** and **spontaneous**. Houses are basic and are built using **cheap materials** that can be found easily, such as plastic, wood and corrugated iron. Squatter settlements have **few services**.

4 As **more and more people** arrive in the city each day, greater **pressure** is placed on the city and its **resources**.

Remember
All LEDC cities are different and we can only talk about general patterns.

Overcrowding: LEDC cities often have a very high population density, particularly on the edge of the city.

Over-population: The growing populations of most LEDC cities have put pressure on services and resources such as clean water, healthcare, transport and housing.

Competition for land: Job opportunities and services tend to be poor in the squatter settlements, so migrants to the city want to live near to the CBD or good transport links. This creates competition for the best areas of land.

Disease: Poor sanitation, limited healthcare and little clean water in squatter settlements means there is a high risk of disease.

Lack of space: Settlements become dangerous if they are built on slopes at risk from mudslides, or if they are built close to factories and industrial areas.

B Case study – Nairobi

Life in Nairobi

Twenty per cent of Kenya's population live in towns and cities. Nairobi, the capital, has over 2 million people. The city has an international airport and good road and rail links to Mombasa on the coast and Uganda to the west. It is a modern city, with cinemas, theatres, expensive shops, museums and government buildings. Housing ranges from luxury apartments to slums.

Half of Nairobi's population live in squatter settlements. Kariobangi is a squatter settlement on the outskirts of the city. It covers an area of about 10 sq km. Sixty thousand people live in Kariobangi. It has no sewerage system, running water, electricity or health centres. There are only three schools in Kariobangi. Open sewers are often blocked with rubbish. When it rains, they overflow leading to health problems. Diseases such as cholera and typhoid are common.

Q Do you know what a squatter settlement is?

C Improving quality of life in LEDC cities

KEY FACT

1 A lack of money and resources means that governments and local authorities in many LEDCs are often unable to do much to reduce the problems created by rapid urbanisation.

2 In the **past,** governments have viewed squatter settlements as a **problem** that must be **cleared away.** Many were literally **bulldozed** out of the way and the residents encouraged, or even **forced,** to go back to the countryside.

Remember
People migrate from the countryside due to push and pull factors.

3 In recent years, **governments** and **charities** have been working with the **communities** that have developed in the squatter settlements to help them **improve the quality of life** for individuals and their families.

Site-and-service schemes give people the chance to rent or buy a piece of land that is connected to the main services of the city (e.g. water and roads). People build their own home on the land, often using money from a loan. They are encouraged to make use of their own skills and ideas to build their homes.

Self-help schemes organised by government or charities encourage the people living in shanty towns to improve their homes. Materials are supplied and loans given. If the area improves, the residents are sometimes given legal ownership of the land. Over time, the services in the area also improve.

Q How can people improve squatter settlements?

As well as trying to improve the quality of life in squatter settlements, governments in many LEDCs have tried to improve living conditions in rural areas. It is hoped that such schemes will discourage people from migrating to the city.

PRACTICE

1 Use the case study of Nairobi above to describe some of the problems faced by those living in squatter settlements.

2 Explain how the quality of life for residents of a squatter settlement may be improved.

THE BARE BONES

➤ The inner city is an area that has experienced a lot of change during the past century.

➤ It is one of the oldest areas of many MEDC cities. In recent years, people have tried to redevelop and regenerate many inner–city areas.

➤ Many MEDC cities were developed before the use of cars and buses, which has led to problems of traffic congestion and pollution.

A Changes in the inner city

KEY FACT

1 The inner city is located <u>close to the CBD</u> and is a zone of <u>traditional industry</u> and <u>nineteenth-century housing</u>.

2 The inner city developed during the **industrial revolution** (1800s), when cities began to grow. Factories and industry developed just outside the CBD and small **Victorian terraced homes** were built close to factories to house the workers.

3 During the **nineteenth century**, large numbers of people moved from the countryside to the city looking for work (urban–rural migration). Migrants tended to find **low-cost housing in the inner city**. The area became densely populated and living conditions were often poor.

4 During the **twentieth century** (1900s), the inner city experienced many changes. Traditional **manufacturing industry** declined and most of the factories closed. Industry moved to the edge of the city. This left the inner city with many problems. A **spiral of decline** was set in motion.

Remember
MEDC cities grew quickly due to rural–urban migration, just like LEDC cities today. Low-cost housing was put up quickly and in high densities, which is why living conditions were poor.

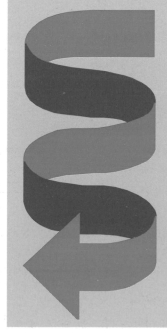

- A loss of jobs in the area led to high unemployment.

- The low-cost Victorian terraced housing fell into disrepair and residents could not afford to improve them. Areas of land were left derelict and the physical environment fell into decline. Vandalism and graffiti became a problem in many areas.

- People moved away from the area. This is known as out-migration.

- Services closed and there were few opportunities for entertainment or leisure. Social problems developed as more people fell into poverty and crime rates increased. Low-income immigrants to the UK tended to find their first home in the inner city. This created social and racial tension.

- Many inner-city areas have experienced unrest and rioting since the 1960s.

Q Can you list the problems faced by many inner city areas?

B Urban regeneration in the UK (an MEDC)

1 After **World War II**, the UK government tried to clear many inner-city areas by demolishing the houses and factories. This was called **slum clearance**. They rehoused people in new high-rise flats or moved them into **council estates** on the edge of town.

KEY FACT

2 In the <u>1980s</u> and <u>90s</u>, the <u>government</u> and <u>private companies</u> tried to <u>redevelop</u> and <u>regenerate</u> many inner-city areas.

✓ **Redevelopment** involves improving the physical environment by clearing old buildings and developing new business offices and housing. It is hoped that this will attract **new business** and **investment** into an area. This helps to create employment opportunities and raise the quality of life. It is known as **regeneration**.

Q Can you name the advantages and disadvantages of regenerating inner cities?

✗ Urban redevelopment and regeneration can create **conflict**. For example, in the **London Docklands**, house prices are now too high for local people. The jobs created by regeneration tend to be IT-based and high-skilled. They are taken by young, well-qualified workers from outside the Docklands area. Hostility has developed between the original residents and newcomers to the area.

C Transport problems in cities

1 Many MEDC cities were developed before the widespread use of **cars and buses** and were not designed for the heavy use of cars and other vehicles.

2 Many people now live on the edge of cities and **commute** (travel) to the city each day for work.

Remember
Make sure you understand the difference between redevelopment and regeneration. To help improve an area you usually need to do both.

3 As cities have grown outwards and urban sprawl has taken place, **large roads and motorways** have been built on the edge of many cities. These link up with other smaller roads that bring cars into the city centre. Many of the roads in the inner parts of the city tend to be narrow and in need of **regular maintenance**.

4 Most cities are major **route centres**. This means that several different routes (**road and rail**) all converge (meet) in the city. This has created problems of **congestion and pollution**.

Q Do you know why many cities suffer from traffic problems?

5 During recent years, the government and local councils have tried to reduce the problems of traffic congestion in cities. **Traffic management schemes** include **improving public transport** and making it more integrated, developing **park-and-ride schemes**, encouraging people to cycle by creating **cycle lanes** and building new roads. Some **roads are made narrower** to restrict traffic.

PRACTICE

1 Explain why inner-city areas in MEDC cities are often areas of decline.

2 Explain how the inner city can be improved.

3 Using examples, describe how traffic problems can be managed in urban areas.

The urban-rural fringe

THE BARE BONES

➤ The urban-rural fringe is the point at which the city meets the countryside.

➤ Green belts have been established around many towns and cities to prevent urban sprawl.

➤ Many people are moving from cities to the countryside. This is called counter-urbanisation.

A The green belt

1 The area where the city meets the countryside is called the **urban-rural fringe**.

KEY FACT

2 MEDC cities tend to grow outwards. The spread of urban areas into the countryside is called <u>urban sprawl</u>.

3 As an urban area grows, it can merge with smaller towns and villages found around the edge of the city. This leads to the development of **urban conurbations**.

4 In the UK, **green belts** have been established around many large cities to try and prevent urban sprawl. A green belt is an area of **green land around the city** where **development is restricted** and controlled. It is designed to restrict the growth of a city and to protect the countryside.

5 **Conflict** can arise over how the land on the green belt is used. There is growing pressure to develop areas of green-belt land.

Q Can you explain what is meant by 'urban sprawl'?

B Counter-urbanisation

KEY FACT

1 The movement of people <u>out of urban areas</u> to live in rural areas is called <u>counter-urbanisation</u>. This is <u>urban to rural migration</u>.

2 Counter-urbanisation is a process happening **in many MEDCs**.

3 People want to live in **open areas** surrounded by greenery, away from the bustle and noise of urban areas. **Houses** tend to be **cheaper** in rural areas, although as the demand increases for rural homes, the price is rising.

4 **Improvements** in transport have made it possible for people to **live** in the **country** and **commute** to work in the **city** each day. This has led to the development of **commuter villages**, where many of the residents **do not** work in the village or surrounding area.

5 Improvements in **telecommunications** also make it possible for more people to **work from home**. This is called **teleworking**.

6 Counter-urbanisation has meant that the **population** of many **inner-urban areas** has **declined** over the last thirty to forty years.

Q Can you list the attractions of migrating from urban areas to rural areas?

C Out-of-town shopping

1 **Traditionally**, most of the **services** found in a town or city were located in the **centre** of the settlement.

2 The decline of the inner city and the migration of people to the suburbs has meant that many **services have relocated to the edge** of the city. This has led to the growth of out-of-town shopping centres. There is **more land** available at the edge of cities and at a **lower cost** than in the CBD. The use of **greenfield sites** on the edge of the city means that there is room for expansion in the future.

3 The movement of services from the CBD has been encouraged by the rapid rise in the number of people owning **cars**.

✓ Most out-of-town shopping centres are located close to or alongside the **motorways** that surround the city. People can get to them easily and the availability of land means that retailers can provide plenty of **free car-parking**.

✓ In recent years, large **retail parks** and regional shopping centres have been developed on the edge of cities. They provide a **large range of stores in one location** and have the same range of high-order shops found in the centre of a town or city, yet they are often easier to get to and have longer opening hours.

✗ Out-of-town centres provide a convenient way to shop, but they **threaten shop owners in the CBD**, who may lose money due to a decline in trade.

✗ **Environmental groups** often oppose new developments on the edge of cities, because these projects mean a **loss of green, open land** and encourage the use of cars, which **pollute** the atmosphere.

Bluewater Shopping Centre, Kent

Remember
The fact that most people in MEDCs now own cars has changed urban patterns, including where people live, work and shop.

Q Can you explain why the development of out-of-town shopping centres can cause conflict?

PRACTICE

1 Explain why green belts were introduced around many UK cities.

2 Explain what is meant by counter-urbanisation.

3 Why is the urban-rural fringe an attractive location for many out-of-town shopping centres?

Make sure you can name an out-of-town shopping centre and describe its location.

THE BARE BONES

➤ There are four main sectors of industry: primary, secondary, tertiary and quaternary.

➤ The numbers of people working in each sector varies between countries and over time.

A Sectors of industry

KEY FACTS

1 <u>Industry</u> is any economic activity that involves collecting raw materials, producing goods and providing services.

2 <u>Employment</u> is the job people do in an industry.

3 There are **four** main sectors of industry:

Remember
All four sectors of industry are interlinked.

Primary industry collects raw materials by growing, catching or extraction. The main industries include fishing, forestry, mining and quarrying. A job would be a **miner extracting bauxite**.

Tertiary industry provides services. The main industries include retailing, education, healthcare, administration and transport. A tertiary job would be a **salesman selling a car** in a showroom.

Industrial sectors

Q Can you write your own example of jobs for each sector?

Secondary industry processes raw materials and **assembles a finished product**. The main industries include metal making, assembling machines and vehicles and construction. A secondary job would be a **foundry worker processing bauxite** into aluminium and a **factory worker assembling the aluminium** in a car body.

Quaternary industry researches and develops new products or provides new services. The main industries include marketing, advertising, telecommunications, biotechnology and information technology. A quaternary job would be a **design engineer** researching and developing new car models.

B Employment structure

1 Employment structure is the percentage of workers in primary, secondary and tertiary industry in a given area. It can be shown as:

Remember
Employment structure is not fixed; it constantly changes.

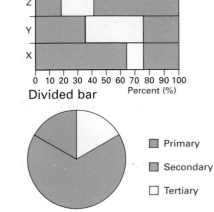

Divided bar

Pie chart for country X

- Primary
- Secondary
- Tertiary

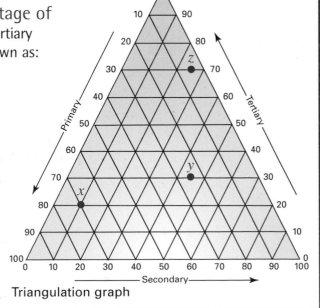

Triangulation graph

B

2 Employment structure **varies** between countries.

- **MEDCs** have a high percentage employed in **tertiary** industry.
- **LEDCs** have a high percentage employed in **primary** industry.

3 Employment structure also varies in a country over time.

- **Pre-industrialisation**: a high percentage work in **primary** industry.
- **Industrialisation**: a significant percentage work in **secondary** industry.
- **De-industrialisation**: a high percentage work in **tertiary** and **quaternary** industries. The UK is now in this phase.

Q Does employment structure vary between MEDCs and LEDCs?

C *Industry as a system*

1 Each sector and type of industry within that sector can be studied as a **system**. There are three main components:

- **Inputs:** What is needed to make the goods.
- **Processes:** Ways of making the goods.
- **Outputs:** What is produced for sale, profits and waste products.

Inputs: Raw materials, labour, energy, capital, machinery, buildings, packaging.

Processes: Assembly, processing, packaging, administration, maintenance, advertising.

Feedback: Reinvestment of some profits.

Outputs: Finished products, waste products, profits not being reinvested.

Q What are the main outputs in any industrial system?

PRACTICE

1 a) Divide the following list into jobs and industries: teacher, car assembly, doctor, web design, train driver, builder, health, education, logger, construction, retail, scientist, forestry, jeweller, manufacturing, information technology, research and development, transport.
b) Subdivide the list into primary, secondary, tertiary and quaternary jobs and industries.

2 Give the employment structure breakdown for countries X, Y and Z using the triangulation graph opposite.
a) Which is an LEDC?
b) Which suggests de-industrialisation?

3 Complete a systems diagram for the car industry (the industrial sectors spider diagram will help).

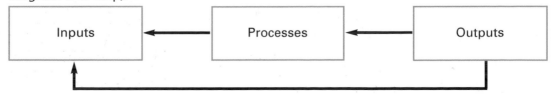

THE BARE BONES
> Agriculture is the growing of crops and the rearing of animals.
> There are many different types of agriculture.

A Farming as a system

KEY FACT

1 As farming is a <u>primary industry</u> with inputs, processes and outputs, a <u>systems diagram</u> can be used to explain it.

2 The system can be used for **all types of farms** in LEDCs as well as MEDCs.

Remember
The farmer is the decision-maker in the system.

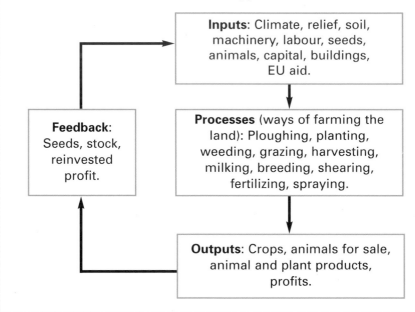

Inputs: Climate, relief, soil, machinery, labour, seeds, animals, capital, buildings, EU aid.

Feedback: Seeds, stock, reinvested profit.

Processes (ways of farming the land): Ploughing, planting, weeding, grazing, harvesting, milking, breeding, shearing, fertilizing, spraying.

Outputs: Crops, animals for sale, animal and plant products, profits.

Q What are outputs?

B Classifying farms

KEY FACT

There are <u>three</u> main ways of classifying farming.

Remember
Farms can be classified in a variety of ways.

1 Farms may be classified by **what** they produce.

- **Arable** farms grow crops.
- **Pastoral** farms rear livestock.
- **Mixed** farms do both.

Pastoral farm

Arable farm

B

2 Farms may be classified by **why** they produce.

- **Subsistence farming** is when animals or crops are grown for the use of the farmer and his family. Any produce that is left over (surplus) is sold at a local market to give more income to buy other goods.

- **Commercial farming** is when animals or crops are grown for sale by the farmer. The farmer gets his income from the profits he makes at market.

3 Farms may be classified by the **level of input**.

- **Extensive farms** are large and have low inputs of labour or technology. Yields per hectare are low. The size of the farm guarantees sufficient produce for survival.

- **Intensive farms** are smaller and have high inputs of labour or technology. Yields per hectare are high.

4 Farms may be classified by their **permanence**.

- Farming is **nomadic** where the farmers have to move their homes or animals regularly to maintain production.

- Farming is **sedentary** where the farmers do not have to move and can remain fixed in a single location.

Nomadic farming

Q Can you think of a farm type for each classification?

PRACTICE

1 List the inputs for a rice farm in India.

2 List the processes for a sheep farm in Wales.

3 List the outputs for a dairy farm in England.

4 Give an example of a nomadic type of farming.

5 What is an extensive farm?

Questions on farming often focus on systems diagrams. Practise systems diagrams for different types of farm (e.g. arable or hill farming).

THE BARE BONES
➤ Farming is affected by human and physical factors.
➤ Farming patterns are affected by climate and relief.

A Factors that affect farming

1 A farmer has to decide which **type** of farming will be most **appropriate** for the **location** of his farm, taking into account several factors.

2 These can be **physical**.

Is the relief suitable?
Steep land is unsuitable for machinery. Arable farming needs fairly flat land. Southern aspects (the way the land faces) are best for crops. Temperatures fall by 0.6 °C every 100m uphill.

Physical factors a farmer must consider

Are the soils suitable?
Crops need well-drained land. Crops need deep, fertile soils.

Is the climate suitable?
Most crops need a growing season of 60+ days of temperatures greater than 6 °C. Most crops need regular annual rainfall between 250–1000mm. Dairy farms need higher levels of rainfall for good grass.

3 They can be **human**.

Are lots of people or technology needed?

What EU government help is there? Subsidies may be available. There may be guaranteed minimum prices. There may be quotas imposed.

Human factors a farmer must consider

How accessible is the farm to a market?

Is the market local, national or global?

Remember
Most farmers want to make a profit.

Q Why is relief important to farmers?

B Farming types in the UK

KEY FACT

1 Farming in the UK is heavily influenced by the physical factors of <u>climate</u> and <u>relief</u>.

- Most **arable farming** and **market gardening** is in the **east** and **south**.
- Most **livestock farming** is found in the **west** and **north**.

KEY FACT

2 Without EU assistance, farming would be unprofitable and could not continue in parts of the UK.

C Distribution of farming in UK

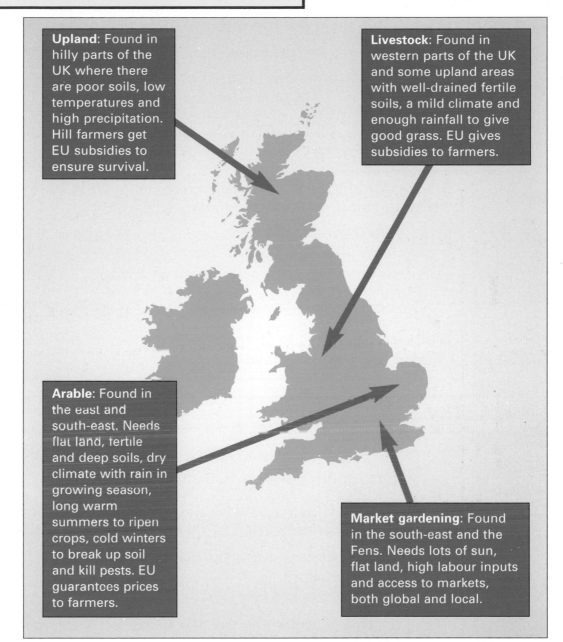

Upland: Found in hilly parts of the UK where there are poor soils, low temperatures and high precipitation. Hill farmers get EU subsidies to ensure survival.

Livestock: Found in western parts of the UK and some upland areas with well-drained fertile soils, a mild climate and enough rainfall to give good grass. EU gives subsidies to farmers.

Arable: Found in the east and south-east. Needs flat land, fertile and deep soils, dry climate with rain in growing season, long warm summers to ripen crops, cold winters to break up soil and kill pests. EU guarantees prices to farmers.

Market gardening: Found in the south-east and the Fens. Needs lots of sun, flat land, high labour inputs and access to markets, both global and local.

Remember
Upland farming means sheep.

Q What affects farming patterns?

PRACTICE

1 Explain three physical factors that a UK farmer must consider.

2 Explain three human factors that a UK farmer must consider.

3 Describe the characteristics of an arable farm.

Be able to describe the UK's distribution of farming.

Changes in MEDC farming

THE BARE BONES

➤ In MEDCs there are fewer farms now than in the 1950s.
➤ Today's farms are generally bigger and more productive.
➤ Changes to farms have caused economic and environmental problems.

A Changing farms

Remember
MEDC farming has changed since 1950.

1 After **World War II**, most MEDCs needed to **increase agricultural output** to feed **growing populations**. This was achieved by:

- creating bigger farms
- draining wetlands to give more land
- more specialisation (e.g. growing only one or two crops)
- building large stores for produce
- making bigger fields by removing hedgerows and trees
- using more machines and fewer people
- using more fertilisers, herbicides and pesticides

Q What is 'intensification'?

2 This is called **intensification**. Intensification has resulted in many changes in the landscape of the countryside.

B The European Union

KEY FACT

1 The fifteen countries that belong to the EU apply the Common Agricultural Policy to all farms.

- The **CAP** aims to provide EU residents with supplies of food at **reasonable prices** while also **protecting** EU **farmers** from foreign competition.
- The **CAP** guarantees farmers a **standard price** for their products.

KEY FACT

2 Britain joined the EU in 1973.

- Like farmers in other member countries, UK farmers were producing **high yields**. The standard price was encouraging farmers to produce more and more food. Produce had to be held in **storage** until it could be sold at the standard price.

Remember
CAP is the Common Agricultural Policy.

3 **Over-production** had given the EU a massive **surplus** of food and wine, known as the **grain mountain** and the **wine lake**.

4 In the **1980s**, the EU introduced policies to **reduce surpluses**.

- In 1984, **dairy farmers** were given **quotas** limiting the amount of milk they would be paid to produce.

- **Arable farmers** had to apply **set-aside** to their land.

Permanent set-aside means that farmers leave 15% of their land uncultivated. They are paid £250 per hectare, per year to do this. They may leave this land fallow (covered with grass), plant woodland, set up wildlife areas or use the land for business that is not farming.

B

5 **Diversification** is using land for non-farming purposes. Some farmers, especially those close to towns or areas of scenic beauty, have found **diversification more profitable** than farming and have taken more of their land out of production.

Setting up craft centers/tearooms · Develop golf course · Pony trekking/ riding schools

Forms of diversification

Sell land for development · Campsites/B and B/ holiday cottages · Farm products/ shop, farm visits/ attractions

Remember
Britain belongs to the EU.

6 The EU continued to encourage lower levels of production in the 1990s by introducing subsidies to farmers in environmentally sensitive areas to farm in an environmentally friendly way.

- This has helped farmers to meet the growing demand for **organic produce** grown without artificial fertilisers, herbicides or pesticides. The subsidies make up for the higher costs of sustainable farming and the lower yields of organic farming.

Q What is 'diversification'?

C *Environmental change*

KEY FACT

Intensive methods of farming in MEDCs have brought many environmental problems.

- Wetlands drained to increase areas in production reduces water supplies and destroys unique insect and bird habitats.
- Hedgerows removed – destroys local ecosystems and takes away wildlife that would naturally control pests.
- The larger fields are more vulnerable to soil erosion once binding root systems have been removed.
- The overuse of fertilisers sprayed on fields washes nitrates into lakes and rivers, causing plants and algae to grow. These use up the available oxygen supply in the water, suffocating other plants and fish. This is eutrophication.
- Increased production has led to large quantities of some natural waste products, such as slurry from animals and ammonia from silage, reaching water supplies. This has a similar effect to the overuse of fertilisers and can cause Blue Baby Syndrome.

Q Can you describe 'eutrophication'?

 **PRACTICE**

1 Write down four differences in the countryside between 1930 and 2000.
2 What are the aims of CAP?
3 How has the EU overcome over-production?
4 Why are nitrates bad for the environment?

THE BARE BONES

➤ Most people in LEDCs work on farms and are subsistence farmers.

➤ LEDCs need to increase food output to meet the needs of their rapidly growing populations.

A Farming types

KEY FACT

1 Most farming in LEDCs is <u>subsistence</u>.

Remember
LEDCs have different farming needs.

2 Subsistence farmers can easily fall into the **property trap**. Yields are low and a lack of money means that **few technological advances** can be introduced to improve them, so there is never money to invest in improvements. The most common types of subsistence farming are:

- **Intensive arable**, which uses large amounts of labour on small plots of land, such as for **rice farming** in the **Ganges Delta**.

- **Intensive shifting cultivation**, which uses large amounts of labour in small clearings in forests, as practiced in the **Amazon Basin**.

- **Extensive pastoral nomadism**, which uses small amounts of labour and vast amounts of land, as practised by herdsman seeking out grazing land for animals in the **Sahel** region of Africa.

KEY FACT

3 There is some <u>commercial</u> farming in LEDCs.

4 Most commercial farms are **very large** in size, are **run by MEDC companies** and are the legacy of a **colonial past** (see 'Development' p.118).

- **Plantations** use high inputs of technology and labour to produce a single crop (e.g. rubber or bananas) in places such as the **Caribbean**. This is **mono-culture**.

- **Ranches** use high inputs of technology to clear land and raise animals for meat products in places such as **Brazil**.

KEY FACT

5 Commercial farming provides both <u>costs</u> and <u>benefits</u> to LEDCs.

 **Benefits**

- regular employment
- workers skilled in using modern machinery and techniques
- export earnings for LEDCs

 Costs

- low-paid employment
- profits go back to MEDCs
- over-reliance on a single crop makes LEDCs vulnerable to weather, disease, and a drop in demand for the crop
- environmental damage caused to large areas of land

Q Can you explain 'mono-culture'?

B Changes in farming

1 In many LEDCs, <u>farming yields</u> have not increased at the same rate as the country's <u>population</u> over the last fifty years.

- There is increased pressure on the land to provide **food**. Marginal land is now being cultivated for production. Over time this causes **land degradation**.

2 In the **1960s**, LEDCs used capital-intensive techniques to improve food production. This is known as the **Green Revolution**.

3 MEDC scientists developed new strains of rice, maize and wheat. These **high-yield varieties of crops (HYVs)** were sold to farmers across South America and Asia. **Crop yields trebled**, because the new crops had shorter growing seasons and so several crops could be harvested in a year. Also, the crops were smaller, allowing more plants to be grown closer together.

4 However, the <u>Green Revolution</u> has been only partially successful, creating costs as well as <u>benefits</u> for LEDCs.

 Benefits

Increased yields, large farmers improve living standards, wider variety of crops, foreign exchange earned from export of crops, spin-offs in the local economy, better transport systems, locals gain skills to make fertilisers and pesticides, knowledge of irrigation systems.

 Costs

Poorer farmers cannot afford HYVs, unemployment as farm machines replace people and small farmers are bought out, increased rural-to-urban migration, crop prices fall as yields increase so small farmers get no benefit, many crops exported, no improved local diet.

C Farming in the future

1 LEDCs have now learnt that MEDC farming methods do not always work for them.

2 <u>Appropriate technology</u> is now being applied in LEDCs to raise production in a way that is sustainable for both people and the environment.

3 To produce enough suitable food for all the people at reasonable prices, LEDCs are:
- developing **simple equipment** which can be easily maintained
- using **affordable local labour**, skills and materials so farmers avoid debt
- using farming methods that use **natural fertilisers** and do not harm the land.

1 What types of farming are there in LEDCs?

2 Why do LEDCs need to increase food production?

THE BARE BONES
➤ The types of industry found in a country change over time.
➤ The location of industry is influenced by four main factors: raw materials, transport, labour and markets.

A The location of industry

KEY FACT

1 Usually a <u>combination of factors</u> affect the location of industry.

Transport costs · Capital · Fuel supply · Decision-makers · Government change · Linked industries · **Location factors** · Labour supply · Agglomeration economics · Globalisation · Raw materials · Proximity to markets · Technological change

Remember
Decision-makers choose the location.

Q Which location factors are important to heavy industries?

2 Once a general area has been chosen for an industry, the exact location must be found. **Local site factors**, such as size of site and transport links, as well as affordability, become important.

KEY FACT

3 The importance of different factors will vary over time as countries pass through <u>phases of development</u> and the type of industries change.

B Changing locations

KEY FACT

1 Heavy industries such as <u>steel-making</u> and <u>textile production</u> were the first to develop in the UK during the <u>Industrial Revolution</u> in the <u>1800s</u>.

Remember
Industrial location is constantly changing.

- They depended on **bulky raw materials** and energy such as coal and iron. They produced **bulky products** that could be transported to the market.

- Early industry was also **labour-intensive** and needed large numbers of local workers.

- Later, as sources of raw materials ran out, some heavy industry relocated to **ports**, as it was cheaper to process bulky raw materials at the point of unloading rather than inland (e.g. the iron and steel works at **Port Talbot**).

Q Name a heavy industry.

2 Heavy industry has <u>declined</u> in the UK.

- This process of **de-industrialisation** has been especially noticeable since the **1970s**, due to the growth of **multi-national companies** (MNCs) and globalisation. It has had massive economic and social consequences for the traditional regions of industry in the UK.

3 **Globalisation** is the process in which national economies become integrated into a **single global economy**. MNCs are the main force behind globalisation, as 40 per cent of the world's trade is carried out by the 350 largest MNCs, such as IBM and General Motors.

> An MNC is a company that operates in more than one country.
> MNCs design, produce and market goods on a global scale, not a national scale. They will source materials and make goods in the cheapest locations in the world to maximise profits.
> In the UK, both raw materials and labour costs are high. This has made most UK heavy industry uncompetitive in the global marketplace. Lower wages abroad mean goods can be produced cheaply elsewhere.
> Many MNCs have withdrawn or are scaling down their existing UK operations. Ford stopped car production at Dagenham in Essex in 2002 for these reasons, after eighty years of production.

4 **Light industries** such as food processing, high technology and electronic consumer goods have been developed in the UK since the **1950s**.

- Both the raw materials and the finished products are **less bulky** and less reliant on traditional location factors.

- Modern industries are called **footloose**, which means they can locate where they want to.

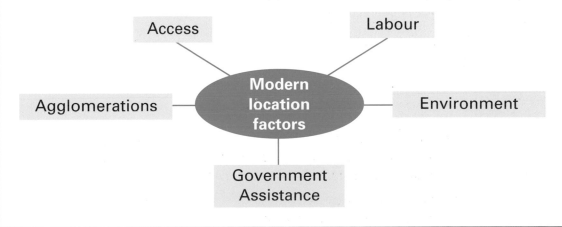

Q What is globalisation?

1 What is 'footloose' industry?

2 Why has heavy industry declined in the UK?

3 Why are MNCs symbols of globalisation?

Modern industry in MEDCs

➤ Economic activity today is global.
➤ High-technology industries make a major contribution to economic output, employment and wealth in many countries.

A High-technology industry

KEY FACT

1 High-technology industry makes <u>high-value, sophisticated products</u>, ranging from biotechnology to consumer electronics (e.g. TVs).

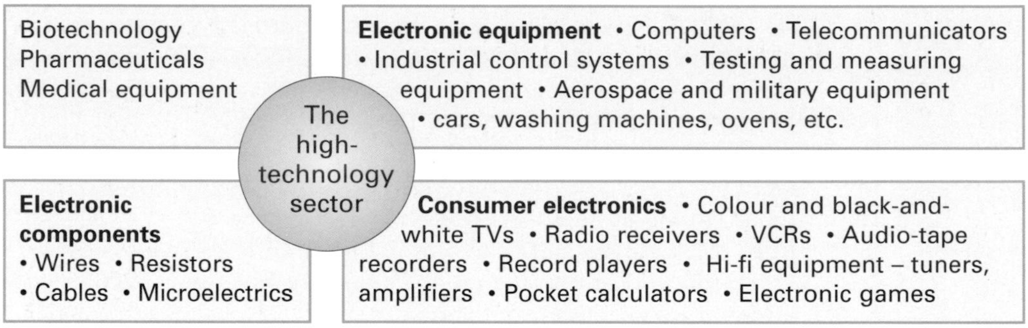

Biotechnology
Pharmaceuticals
Medical equipment

Electronic equipment • Computers • Telecommunicators • Industrial control systems • Testing and measuring equipment • Aerospace and military equipment • cars, washing machines, ovens, etc.

The high-technology sector

Electronic components
• Wires • Resistors • Cables • Microelectrics

Consumer electronics • Colour and black-and-white TVs • Radio receivers • VCRs • Audio-tape recorders • Record players • Hi-fi equipment – tuners, amplifiers • Pocket calculators • Electronic games

KEY FACTS

2 Today's industry is <u>global</u>, with production, organisation and distribution taking place in several countries.

3 Many of the industries in the high-technology sector are operated by large <u>MNCs</u>. Examples of such companies include Sony, Samsung, Marconi and Microsoft.

4 The high-technology sector is a good example of a **footloose** industry. Factories locate where costs are low and profits optimised.

Important location factors for decision-makers include:

- attractive locations for **headquarters and research and development** to attract skilled, highly qualified labour.

- **greenfield sites** for mass **production** and assembly.

- **cheap unskilled labour** for **assembly** plants.

- close to large **markets** as finished products are sometimes bulky.

- nearby links to associated industries (**agglomeration**).

- **government incentives** (low rents, taxes).

Remember
Most high-tech firms are MNCs.

Q What is high-tech industry?

B Case study – Lucky Goldstar (LG)

Who are LG?

LG is a Korean MNC that manufactures electronic components and consumer electronics. It has low-cost production bases located across the world.

In 1997, LG invested £1.7 billion in two factories at Newport in South Wales. These produce semi-conductors and colour tubes for TVs and PC monitors.

Why choose Newport?

Government incentives were available to LG if they chose to locate in Newport. The region is a development area, so it gets financial help from the UK government and the EU. This is because South Wales suffered de-industrialisation in the 1980s.

Greenfield sites were available for development. The Welsh Development Agency made it an attractive environment with improvement of derelict land and the scars left from mining.

Good transport with the M4, linking Newport to associated industries and experts in Oxford and London, and the M5, opening up UK domestic markets.

A linked industry already in the region is Sony at Bridgend, also in South Wales.

High unemployment means a large cheap supply of labour for assembly.

Benefits of LG

LG has been extremely good for the South Wales economy and for rebuilding local communities destroyed by de-industrialisation. These positive benefits are called the multiplier effect:

- Large numbers of construction workers employed to build factories.

- Two hundred core workers received training in Korea.

- A further six hundred people were trained nearby at Cumbran.

- Training and scholarships provided for young people.

- Links made with University of Wales, including research contracts and language courses for Korean staff.

- Agglomeration economies achieved with four companies based in South Wales supplying LG with components worth £4.6 million: mouldings for PC monitors at Llantrisant, polystyrene packaging at Tonypandy, cardboard boxes at Newport, printing manuals at Pontypool.

Remember
LG is a Korean MNC.

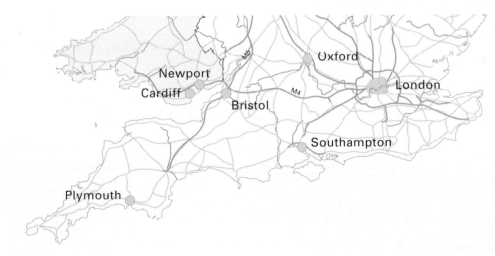

Q Can you name another MNC?

PRACTICE

1 What are the most important factors affecting the location of high-technology industry?

2 How does the government attract new industries to the UK?

3 How has LG regenerated Newport?

Learn the meanings of key terms like 'multiplier effect', 'high-technology', and 'de-industrialisation'.

THE BARE BONES
➤ Most LEDCs are in the pre-industrialisation phase.
➤ A few LEDCs have developed industrialised economies since the 1950s.

A Classifying industry in LEDCs

KEY FACT

1 Some people work in the <u>formal industry</u> sector.

• This is a job with a **permanent** contract in a **factory** or **office**. Workers are employed and paid a **regular wage**. There are more workers than jobs available.

KEY FACT

2 Most people work in the <u>informal industry</u> sector.

• Jobs are **temporary** and insecure and usually based in small **workshops** or on the **streets**. Workers are **self-employed** and do not get regular income.

3 This pattern of industry is typical of LEDCs that have not experienced industrialisation. Most LEDCs will remain in this **phase of development** for some time.

Q Describe a job in the informal sector.

B Newly Industrialising Countries (NICs)

KEY FACT

1 NICs are LEDCs that have seen <u>heavy industrialisation</u> since the 1950s.

2 Early NICs (Brazil, Mexico, South Korea, Hong Kong, Singapore, Taiwan) were concentrated in **South America** and **South East Asia**.

3 New NICs (Thailand, Malaysia, Philippines, Indonesia, China) continue to develop around the **Pacific Rim**.

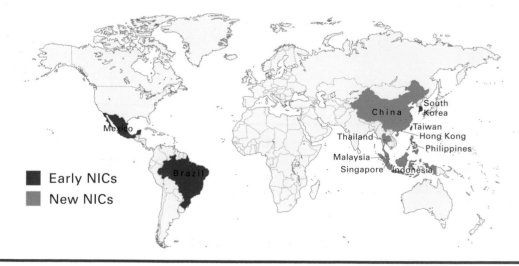

■ Early NICs
■ New NICs

Q Name an early NIC and a new NIC.

C   Achieving industrialisation

1 Some LEDC governments have exploited their large populations (plentiful <u>cheap labour and large domestic markets</u>) and <u>industrialised</u> their societies.

2 Most NICs have taken a series of steps to achieve their aims:

Becoming an NIC

Borrow to develop infrastructure and educate. This attracts MNC's to the country. Use income and skills from MNCs to develop own high-value consumer goods. Import restrictions (to protect domestic markets) devalue currency, making exports cheaper.

Reliance on MEDC MNC's decreases as they seek new markets elsewhere. Industrialisation and NIC status achieved. Further investment in infrastructure and own companies – some become MNC's like LG in South Korea. Living standards improve rapidly, but so do production costs.

Remember
LEDCs need MNCs to develop.

3   MNCs are the agents of NIC development.

• MNCs are attracted to countries such as **Thailand**, with its cheap raw materials, **incentives** (such as low taxes), plentiful, cheap, skilled labour and few trade barriers, giving easy access to large, unexposed consumer markets.

4   MNCs' involvement has both <u>costs</u> and <u>benefits</u> for LEDCs.

✓ Benefits of MNCs	✗ Costs of MNCs
• Often provide initial investment in a country • Speed the development process • Create large-scale employment in the formal sector • Linked industries develop • Skills transfer to host country • Leads to increased investment from outside (multiplier effect) • Spin-offs from increased living standards	• Profits leak back to MEDCs • Much employment is low paid and unskilled • Often poor working conditions • Factories pollute the environment • LEDCs remain vulnerable if MNCs pull out • Westernisation, MEDC values and consumer culture dominate local religious and cultural values

Q Why do MNCs locate in LEDCs?

 **PRACTICE**

1 How are the formal and informal sectors different?

2 Why do LEDCs find it hard to industrialise?

3 What is an NIC?

4 What are the advantages of MNCs to LEDCs?

Tourism

THE BARE BONES
- ➤ Tourism is a tertiary activity and a service industry.
- ➤ Tourism is the world's fastest growing industry employing 10 per cent of the global workforce.

A Tourism as an industry

KEY FACT

1 Tourism is an activity that involves a <u>visit away from home</u>.

- A **tourist** is someone who spends at least one night away from their normal place of residence.
- The visit may be a **holiday** or it could be **business** travel, visiting friends and **relatives**, a religious **pilgrimage** or a trip to gain **health** treatment.

KEY FACT

2 Tourism is a <u>worldwide industry</u> employing 10 per cent of all people of working age.

- Globally, 635 million international visits are made each year.

KEY FACT

3 Tourism has been the world's <u>fastest growing industry</u> since the 1950s.

- People are taking **more holidays**, for longer, and spending more on them.

Remember
Tourism is the world's fastest growing industry.

Changing transport
Improved roads and increased car ownership, introduction of large jet aircraft, more airports.

Changing costs
Travel companies offer cheap, organised package holidays, airlines offer 'no frills' and discounted fares.

Reasons for the growth in tourism

Changing work patterns
Fewer hours worked so more leisure time, more flexible hours, paid holiday, longer periods of holiday leave, increased wages relative to inflation.

Changing attitudes
Public exposed to exotic locations in the media, people now expect to have at least one holiday a year, the active elderly have the time and money to spend on travel, people expect excitement and adventure from holidays.

Q Who can be a tourist?

B Classifying tourism

1 Tourism is often referred to as a resource-based industry. It depends on a combination of primary and secondary resources.

- **Primary resources** occur naturally:
  Attractive climates (sun/snow)
  Landforms (beaches/mountains/lakes)
  Ecosystems/wildlife
- **Secondary resources** are man-made:
  Infrastructure (accommodation/transport)
  Attractions (monuments/religious/stately homes/industrial/historic/heritage sites)
  Entertainment (sporting events/culture/shopping/theme parks)

Remember
Tourism is difficult to classify.

2 Tourism can also be classified by:
- **Location** (seaside/countryside/urban)
- **Activity** (passive/active)
- **Duration** (length of visit)
- **Distance** travelled (local/international).

Questions will ask you to discuss the costs as well as the benefits of tourism.

Q Explain any three ways of classifying tourism.

C The impacts of tourism

1 Both MEDCs and LEDCs want to develop tourist industries because of the wealth it can generate.

2 However, there are often **hidden costs**:

✓ Benefits (Advantages)	✗ Costs (Disadvantages)
Economic Employment; foreign exchange	**Economic** Jobs are often low-paid, of low status and temporary
Socio-cultural Local cultures and traditions maintained	**Socio-cultural** Local culture and traditions debased
Environmental Fragile sites and landforms can be protected	**Environmental** Fragile ecosystems often permanently damaged

Q Why are countries keen to develop tourism?

1 What changes in lifestyle have contributed to the growth of tourism?

2 How can tourism benefit a country?

THE BARE BONES
➤ Mass tourism began within the UK in the 1800s.
➤ Mass tourism to foreign destinations began in the 1970s.

A Changing UK tourism

KEY FACT

1 Early tourism in the UK was based around <u>coastal resorts</u>.

- The 'tourism life cycle' of the UK began in the **1700s** when the 'elite' visited **spas** and early **seaside** resorts. **Mass tourism** developed in the **1800s** with **railway** connections from cities to major seaside towns such as Blackpool. The heyday of bucket-and-spade holidays was in the **1950s**. Saturation of these resorts plus competition from **southern Europe** (in particular, Spain) lured holidaymakers away from UK resorts in the **1970s**. Some resorts such as Margate have gone into decline, but others such as **Brighton** have survived.

KEY FACT

2 To compete with tourism in other MEDCs, the UK has diversified into <u>recreational and urban tourism</u> activities.

Q How has UK tourism changed?

- Urban and heritage tourism celebrates Britain's **historic, cultural, religious and industrial traditions** in locations such as Bath, Salford and Oxford.
- Recreational tourism is concentrated in **scenic upland areas** of the UK.

B National Parks

KEY FACTS

1 Over 100 million visits a year are made to the UK countryside.

2 <u>National Parks</u> are large areas of scenic countryside protected for use by the public, now and in the future.

Remember
It is important to be able to locate each National Park on a map of the UK.

- Since **1951, twelve National Parks** have been created in **England and Wales** and a further park, Loch Lomond in central Scotland, was designated in 2001. Using major roads, most are within driving distance of Britain's major cities:

  - Northumberland
  - Lake District
  - Peak District
  - Snowdonia
  - Brecon Beacons
  - Pembrokeshire
  - Dartmoor
  - Exmoor
  - New Forest
  - Norfolk Broads
  - Yorkshire Dales
  - North York Moors

Q Name a coastal National Park.

3 The aim was to give the public **unlimited access** to these beautiful environments while **preserving** the landscape and looking after the interests of **local people** and businesses.

c Case study — The Yorkshire Dales National Park

Facts and location

- Located to the north and west of **Leeds** and east of **Lancashire**.
- Within 120 minutes driving distance of **9 million people**.
- Approximately **70km wide** and 60km deep at its widest points.
- Experienced **8.3 million** visitor days in **1998**.
- **18 000 people** live within the national park boundary.

Attractions

- Upland area of spectacular **limestone** scenery.
- Beautiful **wooded valleys** and waterfalls.
- **Historic villages** and monuments.
- **Adventure activities**: rock climbing, potholing.

Issues

- Local residents, farmers and businesses all **conflict** with tourist use of the area, but tourism provides the main source of income for local people.
- **Overcrowding** happens when visitors converge on **honeypots**, such as **Malham** village. A 'honeypot' is a site that is the focus of tourist activity in an area.
- **Ninety per cent of visitors come by car**, congesting roads, reducing air quality, causing noise pollution; car parks are visually intrusive in the landscape and overspill at peak times.
- Visitors **erode footpaths** and create gullies when vegetation-cover is trampled, leaving ugly scars on hillsides. Visitors create **litter**, leave gates open and worry animals.
- The sheer volume of visitors has **changed the character** of Malham, which is now dominated by tourist facilities.

This **reduces** the number of **essential services** available to residents.

- Locals are pushed out as **city-dwellers buy second homes** for occasional use at prices locals can't afford. Out of season, the park seems dead.
- Local **limestone quarries** provide local jobs but spoil the view for tourists.

Making the park sustainable

- A **balance** must be struck between the **needs of visitors** and the needs of **local people** and businesses.
- Improving **public-transport links** to the parks, so people leave their cars at home.
- **Locating car parks wisely**, screening them from view and using local materials for the surface.
- Rebuilding footpaths to **limit erosion**, replant in badly eroded areas, limit public access during wet periods when erosion is worst.
- **Restricting development to honeypots** to keep other villages and beauty spots pristine.
- Providing **affordable housing** for locals.

Q What is a honeypot?

PRACTICE

1 Why did UK tourism need to diversify in the 1970s?

2 Use the location map above to explain why the Yorkshire Dales get so many visitors.

3 What are the three main conflicts in the Yorkshire Dales?

Learn as many facts and figures about your case studies as you can.

THE BARE BONES

➤ LEDCs are now popular tourist destinations.

➤ Tourism plays an important part in helping LEDCs to develop.

➤ LEDCs are constantly striving to achieve a balance between the development and exploitation of their tourist resources.

A Tourism and development

KEY FACT

1 Tourism has become a <u>global industry</u>.

- The world is shrinking due to shorter flying times. **Cheaper travel** has opened up far-away locations in LEDCs for the ordinary traveller.

2 Many LEDCs are **attractive destinations**.

Year-round hot climates

Beautiful beaches and scenery

Different cultures to experience

Relatively cheap holidays

Remember
LEDCs are seeking a balance between development and exploitation.

- **Popular destinations** include Thailand, Mexico, Sri Lanka and Kenya.

3 LEDCs are keen to promote their attractions, as tourism helps to <u>fund development</u> of new facilities.

4 The **benefits** of tourism to LEDCs are mostly **economic**.

- Increased wealth.

- Initial investment starting up the tourist industries in LEDCs by multi-national companies (MNCs) in MEDCs.

- Foreign money from trade can be invested in transport, sewerage, water and electricity systems, schools and hospitals.

- New jobs outside primary industry.

- Positive spin-offs for local businesses that supply the tourism industry, such as handicrafts and food and drink.

- Industry will move into LEDCs that have developed infrastructures, generating more wealth. This is the multiplier effect.

Q Name an LEDC tourist destination.

A

5 The **cost of tourism** to LEDCs is mostly **social and environmental**.

- Countries going into debt building their tourism infrastructure.
- Many jobs are only seasonal, low status and badly paid.
- Most profits 'leak' back to the MNCs in MEDCs that initially invested in tourism in the LEDC.
- Often the multiplier effect does not take off, leaving most of the LEDC poor and under-developed.
- Mass tourism often exploits and destroys local traditions and cultures.
- Large-scale environmental damage occurs to ecosystems such as beaches, reefs and forests, which are irreplaceable. There is often pressure on water supplies and land degradation.

Q Can you explain 'leakage'?

B *Sustainable tourism*

KEY FACT

1 <u>Sustainable tourism</u> uses tourist resources today in a way that doesn't damage them for future visitors and local people.

2 Environmentally friendly, low-impact, low-density tourism is an alternative to mass tourism in most LEDCs. It is a **less damaging way to develop** tourism.

3 The main aims of **ecotourism** are to:

- protect the natural environment
- enable local people to earn money
- enable the community itself to improve local infrastructure and facilities
- enable the local community to choose how their tourism industry develops

- Ecotourism encourages tourists to explore LEDC tourist areas in **small groups**, staying in local accommodation, eating local food and observing local customs and culture. These holidays are **expensive** and try to help **conserve** the local environment. Most of the money is returned to the **local economy**. Countries that have developed ecotourism successfully include Costa Rica and Botswana.

KEY FACT

4 As the projects are <u>small-scale</u>, <u>damage</u> to the LEDC is <u>minimal</u>, but as the destination increases in popularity, it becomes more difficult to operate in a sustainable way.

- Often the lure of large amounts of **foreign exchange** from MEDCs is too great and there is not enough expertise within the LEDC itself to protect and manage its own tourist industry appropriately.

Q Can you explain ecotourism?

You need to be able to discuss the economic, social and environmental impacts of tourism in LEDCs.

PRACTICE

1 Why are LEDCs attractive destinations?

2 Explain the multiplier effect.

3 What is sustainable tourism?

4 List three benefits and three costs of tourism to LEDCs.

THE BARE BONES

➤ Mass tourism has developed along the Mediterranean coastline of Spain.

➤ Tourism developed because of abundant resources and cheap packages.

A Facts and figures

KEY FACTS

1 In just over forty years, Spain has transformed its Mediterranean coastline from a string of sleepy fishing villages to <u>international coastal resorts</u>.

2 The <u>Costa del Sol</u> is located in the south of Spain on the Mediterranean. The number of <u>visitors</u> soared from <u>0.4 million</u> in 1960 to approximately <u>7 million</u> in 2000.

3 Before the 1960s, most jobs were in **farming** or **fishing** and the infrastructure was poor. Living standards were low, and migration of the young, high. The region was **underdeveloped**.

4 Numbers peaked at **7.5 million in 1988**. By the late 1990s, Spanish resorts were in **decline**. Prices were too high, while less-developed, **more exotic locations** became available to the masses.

Remember
Spain is at the rejuvenate or decline stage of the tourism life cycle.

Q Is the Spanish coast a good place for tourism?

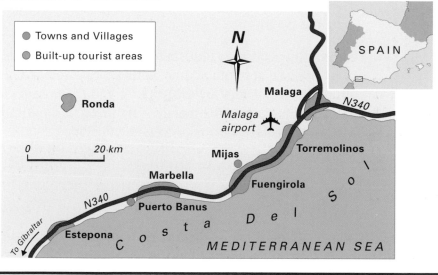

B Tourist resources

Q Can you list Spain's natural tourist resources?

1 Natural resources – dry, warm climate from May to November, long sandy beaches, rugged mountains.

2 Man-made resources – historic and cultural towns, well-developed infrastructure, wide variety of entertainment and nightlife.

C Impacts of tourism

The impact of development along this stretch of coast has been both **positive** and **negative**.

	✓ Benefits	✗ Costs
Economic	• Improved infrastructure (upgrade of electricity supplies, Malaga airport and by-pass). • Improved employment prospects and living standards. • Seventy per cent of people work in tourism.	• Most employment is irregular, low-paid and of low status. • Recently, unemployment in the region has risen to 30%, and many businesses have closed. • Falling house prices.
Sociocultural	• Traditional crafts (e.g. lace-making) and industries have been saved from dying out as goods are sold as souvenirs. • Close-knit communities have stayed intact, cultural traditions (e.g. dances) retained.	• Traditional lifestyles based around the church and family eroded, traditions kept artificially alive, local culture debased into tourist shows. • Increased crime (drugs, vandalism and mugging).
Environmental	• Some areas protected due to investment from tourist revenue. • Some beaches have EU Blue Flag status. • Nature reserves set up.	• High-rise accommodation creates visual pollution. Some 1960s hotels look run-down. • Traffic congestion in towns. • Litter on beaches, sea polluted with sewage from tourist waste. • Pressure on local water supplies in a region with little rainfall.

Q Can you find Mijas on the map?

D The way ahead

1 Due to a **poor media image**, tourists began turning their backs on Spain in the 1990s as **cheaper LEDC holidays** became more available.

2 Steps have been taken to **rejuvenate the Costa del Sol** to attract visitors back all year round.

- Further high-rise development banned. Any new building must be low-rise and in a traditional Spanish courtyard style. Resorts such as Marbella promoted as up-market.
- Resort centres pedestrianised and planted with trees, marinas and by-passes built.
- Development restricted to golf courses and luxury villas between resorts.

Q Why have tourists abandoned Spain?

PRACTICE

1 What are the effects of tourism along the Costa del Sol?

2 How is the region responding to competition?

THE BARE BONES

➤ Most tourism in Kenya has developed along the Indian Ocean coast and in the game parks.

➤ Tourism has brought economic benefits but also social and environmental costs.

A Facts and figures

1 Kenya is on the **east coast of Africa** and was one of the first LEDCs to develop **mass tourism in the 1970s**. It is **English-speaking**, due to past colonial links, and most visitors come from the UK and northern Europe.

KEY FACT

2 In 1997 over <u>$450 million was earned from tourism</u>. This is more than Kenya earns from tea and coffee exports.

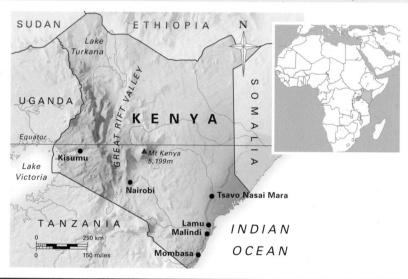

Q Can you locate Kenya on a map of Africa?

B Tourist resources

1 Natural resources – year-round hot climate, sandy beaches, coral reefs, wildlife reserves.

2 Man-made resources – relatively well-developed infrastructure due to colonial links and interesting **diversity of cultures** such as the Masai Mara.

3 Most tourists mix a **beach** holiday at Malindi or Mombassa with a short **safari** on a game reserve.

Remember
Kenya is at the rejuvenate or decline stage of the tourism life cycle.

Q Is Mombassa a good location for tourism?

C Impacts of tourism

The **concentration of tourists** in **environmentally sensitive areas** in Kenya has had a largely **negative** impact on the people and the environment.

	✓ Benefits	✗ Costs
Economic	• 500 000+ employed in tourism. • Improved living standards. • Foreign exchange allows some development of the infrastructure. • Masai can sell firewood to lodges.	• Jobs are often temporary and low-paid. • Foreign MNCs own 80% of hotels and travel companies in Kenya, so profits leak back to MEDCs.
Sociocultural	• Masai settlements used as tourist attractions. • Traditional culture and skills retained.	• Nomadic communities displaced when reserves set up. Loss of dignity and traditional ways. • Alcohol and western flesh offend Muslim coastal communities. • Sex tourism exploits the poor.
Environmental	• The exploitation of a few areas protects the majority of habitats and wildlife. • Profits from tourism can be invested in protecting other environments for the future. • Reserves allow endangered species to thrive.	• Coral reef damaged by people and boats, local fishermen can't use the seas as the ecosystem is destroyed. • Overuse and inappropriate development of the shoreline. • Game park minibuses churn up the bush, create soil erosion, altering the animals' behaviour. • Balloon safaris frighten animals. • Savannah hotels/campsites use limited freshwater and wood.

Q List the economic benefits of tourism.

D The way ahead

KEY FACT

1 Due to violent crime, tourist harassment, civil unrest, <u>over-commercialisation</u> of safaris and destruction of natural resources, <u>visitor numbers dropped</u> steeply in the 1990s.

2 The Kenyan government is now acting to **protect** Kenya's tourist industry by limiting the use of existing marine and game parks and **taxing tourists** and holiday companies who use them. It is also encouraging:

• **Sustainable tourism** on safari. The Tsavo Game Park has temporary camps, limited power supplies and trails in **small groups** that use **local people**.

• Sustainable coastal tourism. At Lamu, visitors pay a local **tax**, stay in small guesthouses and observe **local customs**. There's no development above tree height.

Q Why is Kenya not so popular?

PRACTICE

1 What tourist resources does Kenya offer?

2 How can tourism be socially damaging?

3 How has Kenya managed the impact of tourism?

Development

THE BARE BONES

➤ Development involves improving a country's economy and raising standards of living.

➤ Different indicators can be used to measure a country's level of development.

A What is development?

KEY FACT

1 Some countries are regarded as more developed than others.

- **Wealthier** countries with high standards of living are often called 'more economically developed countries' (**MEDCs**).

- **Poorer** countries with generally low standards of living are often called 'less economically developed countries' (**LEDCs**).

2 Most MEDCs used the process of **industrialisation** to help them develop. The growth of manufacturing industries meant people were able to earn a regular wage and the government spent money on **health**, **education** and **welfare**.

3 In recent decades, many LEDCs have begun the process of industrialisation as a way of developing their economy. However, LEDCs are often held back from development due to **rapid population growth, limited capital for investment** and **debt** (see opposite).

Q Do you know what the terms 'MEDC' and 'LEDC' mean?

B How can we measure development?

1 Deciding how developed a country is can be difficult. Different criteria can be used to measure development. These are called **indicators of development**.

2 In the past, a country's level of development was measured by **assessing its wealth**. This was done by calculating a country's **Gross National Product (GNP)**, which is the amount of wealth produced through trade, services and industry in one year. GNP figures were given in US$ and calculated per person.

3 Recently, governments and international organisations have begun using a **wide range** of indicators to measure development, including social and political criteria:

- **average life expectancy:** how long the average person in a country is expected to live

- **infant mortality:** how many babies die before their first birthday

- **adult literacy:** the number of adults who can read and write

- **access to clean water:** the percentage of the population with access to it

- **calorie intake:** the average number of calories eaten per day

- **access to healthcare:** the number of patients per doctor

- **women's rights:** the percentage of women in government.

Q Can you list five indicators of development?

C The north–south divide

1 The <u>North/South divide</u> is the idea that the world can be split in half: the north and the south. Most MEDCs are found in the north, while most LEDCs are found in the south.

The North/South divide was written about and described in a report called the **Brandt Report** (published in 1980).

- The North/South divide is shown on the map below. The map is not perfect and makes many generalisations:

Remember
Development is not just about money. Social and political conditions are also very important.

All LEDCs and MEDCs are different; try not to generalise about a country. In your exam answers, use terms such as 'generally', 'mostly' and 'tend to'.

- **Australia** is an **MEDC** but it is in the **Southern Hemisphere**. The divide between the north and south has been **modified** to take Australia into account.
- Some **eastern European countries** have very **low levels of economic** and **social development,** yet they are found **north** of the divide.

2 Although the North/South divide is a useful starting point for thinking about differences in development, it is important to understand that differences in wealth exist within individual countries. The **wealth** of a country is **rarely shared** out **equally**.

- The **gap** between the rich and poor countries of the world is **growing**. This is called the **'development gap'**.

Q. What is the development gap?

1 Explain what is meant by the following terms:
a) GNP per capita,
b) adult literacy rate.

2 Using examples, describe the differences in quality of life between LEDCs and MEDCs.

Trade and aid

➤ The world trade system is dominated by MEDCs.

➤ LEDCs do not always benefit from international trade.

➤ Poor terms of trade and donations of aid can make LEDCs dependent on MEDCs.

A Development through trade

KEY FACT

1 Many LEDCs have tried to develop their economies through international trade.

Remember
LEDCs are often disadvantaged by the world trade system. MEDCs tend to benefit from international trade; LEDCs tend to lose out.

2 Trade involves the **exchange of goods and services** within and between countries. Trade between countries is called **international trade**. Importing goods involves buying goods from other countries. Exporting goods involves selling goods to other countries. The **difference** between the **value** of **imports** and **exports** is known as the **balance of trade**.

3 The pattern of world trade is **uneven**. LEDCs have a **small share** of world trade compared to MEDCs. Over 80 per cent of world trade involves MEDCs.

4 Traditionally, most LEDCs have traded with MEDCs by exporting **primary goods** and importing **manufactured goods**. This system of trade favours MEDCs and can cause problems for LEDCs.

Q Can you explain why LEDCs often lose out from international trade?

> **Primary goods**, such as crops (sugar, coffee), are low-value products and do not make large profits for LEDCs.
>
> The value of primary goods can fluctuate a great deal on the world market and LEDCs are not guaranteed a price for their products.
>
> **Manufactured goods** are high-value products. MEDCs make a profit selling them.
>
> **LEDCs** spend more importing manufactured goods than they make selling primary goods. This gives LEDCs a **poor balance of trade**.
>
> **MEDCs** can buy primary products at a low cost from LEDCs and export manufactured goods at a profit. MEDCs have a **good balance of trade**.

B The colonial past

Remember
The past helps us understand the world today. The colonial period still has a big impact on LEDCs.

1 Many LEDCs used to be **colonies**. There were **governed by another country**, usually an MEDC. Many African countries are former colonies. For example, Ghana used to be a British colony and Mozambique was once a Portuguese colony. During the colonial period, colonies (LEDCs) provided their rulers (MEDCs) with a **cheap** and reliable source of **raw materials**. In exchange, MEDCs supplied LEDCs with manufactured goods. This meant that few colonies were given the opportunity to develop their own industries.

B

Q Can you explain why the past has made many LEDCs dependent on MEDCs?

2 Most colonies are now **independent** and not governed by MEDCs. However, their colonial past still affects opportunities for trade and development. Many LEDCs are still dependent on selling primary goods to MEDCs, with **limited industries** of their own. Many LEDCs still need to buy secondary goods from MEDCs.

3 During the 1970s and 1980s, many LEDCs took out **loans** from MEDCs and the **World Bank**. They used the money to try and develop their industries and improve standards of living. Unfortunately, during the **1980s**, interest rates rose rapidly and LEDCs are now left with large **debts** that they cannot repay.

4 The interest owed by many LEDCs is now **more than their original loan**.

C Fair trade

1 **Fair trade** involves making sure that companies and workers in **LEDCs** are **paid a good price** for the goods they produce.

2 **Retailers** (sellers) in MEDCs buy raw materials and manufactured goods from LEDCs at a **guaranteed price** and sell them to consumers. The retailer then makes sure that a **good share of the profits are returned** to LEDCs.

Q Do you know what is meant by 'fair trade'?

3 Fair trade is a **growing** area of business in MEDCs. Many **large food stores** now sell fair-trade items. More and more **product labels** give information about the **origin** of different foodstuffs and the **conditions of the workers**.

D Development through aid

KEY FACT

1 As well as taking out loans from MEDCs, many LEDCs have been given <u>aid to help them develop.</u>

2 There are different types of development aid. Giving aid usually involves **donating money**. However, it can also involve **providing products** and goods to LEDCs.

Bilateral aid: Given directly from the government of one country to another. Includes donations of money, food, technology or training services.

The three main types of aid

Multilateral aid: Donated by several different countries, usually through an international agency or organisation (e.g. the UN or World Bank).

Non-governmental aid: Provided by charities and independent organisations (e.g. Oxfam). It's not directed by governments and often relies on fundraisers and volunteers. Non-governmental organisations (NGOs) tend to support small-scale local development projects (e.g. running health clinics or building a freshwater well).

Q Can you list three types of aid?

3 One of the **biggest concerns** about using aid for development in LEDCs is making sure that the aid given to poorer countries **is appropriate** to their needs and **reaches** the people who need it most.

PRACTICE

1 Explain why LEDCs tend to have a poor balance of trade.

2 Why are LEDCs said to be dependent on MEDCs for trade and aid?

Development projects

THE BARE BONES

➤ Development in LEDCs can involve both large-scale projects and small-scale local projects.

➤ Small-scale local development projects are often better for communities and the environment than larger projects.

A Halving world poverty

KEY FACT

1 1 Over <u>one billion</u> people live in <u>absolute poverty</u>, surviving on just 65p a day.

Q Who has set the target to halve world poverty by the year 2015?

2 The **UN** has set a global target to **halve world poverty by 2015**. Charities, governments and international organisations are involved in meeting this target.

3 Development projects are often used to reduce poverty and improve quality of life in LEDCs. Some are **large-scale projects** and involve **foreign investment**; others are **small-scale** and rely on the knowledge and skills of **local people**.

B Large-scale development projects

1 Many LEDCs use large development projects to try and generate income and improve their country's infrastructure. **Infrastructure** means the **basic services and utilities** needed in an industrial or urban area (e.g. roads, railways, water supplies, medical care). Examples of large-scale development projects include:

Remember
Development in LEDCs can often involve a mixture of large-scale and small-scale development projects.

• Building a **dam** to help provide **water and electricity** in a country (e.g. Three Gorges Dam, China).

• Large **road construction** schemes, such as the Trans-Amazonian highway.

2 Large development projects tend to rely on **heavy foreign investment and aid**. This money often comes from **international organisations**, such as the World Bank.

3 The use of large-scale projects is called **'top-down'** development. This means that a lot of **money is invested** in one large scheme in the hope that the **benefits** of the project will **trickle down** to other parts of the country or region. Development projects can set off the multiplier effect.

Q Can you list the problems involved in large-scale development projects?

• The **multiplier effect** means that when a **large project** is set up, **other jobs** are **generated** as people and services are needed to help with the project. Other **businesses** may be **attracted** to the area and these also **create jobs** and income.

4 Large-scale development projects may create a lot of **problems**. LEDCs often need to borrow money from MEDCs, and this adds to their **foreign debts**. Large projects can **disrupt local communities**. People often lose their homes and have to be resettled.

B

- In recent years large development projects have become **unpopular** with some governments and international organisations. There has been a lot of **publicity** about the **failure of large-scale projects to improve the quality of life for poor communities** and there have been concerns about **corruption**. Organisations like the World Bank are now reluctant to support very large development schemes.

C Small-scale local development

Remember
Small-scale development projects aim to be sustainable and appropriate to the needs of local people.

KEY FACT

1 During recent years, **small-scale development projects** have become a more popular **alternative** to large-scale, top-down development projects.

2 Small-scale development is sometimes called **grassroots development**. This means that development schemes begin by making small changes in an area by **working with local people** and **local skills**. Hopefully the changes will bring about **long-term changes** for an area. This can be seen as **'bottom-up'** development.

3 **Small-scale development projects try to help meet people's basic needs, such as adequate clean water or food.**

4 Grassroots projects **do not depend on** a lot of **heavy investment** from foreign companies. Instead, local people are encouraged to raise **their own money** or take out small loans from **local banks**.

5 **Non-governmental organisations (NGOs)** often provide **training** and **education** to help communities with grassroots projects.

Case study – grassroots development in Nepal

Intermediate Technology is a charity involved in small-scale development in a number of different LEDCs. One of the projects run by the charity is based in rural areas of Nepal.
Deforestation is a problem in Nepal, as people use firewood as fuel for cooking. Intermediate Technology has been teaching local women to make fuel-efficient cooking stoves. These stoves use less firewood, which helps to protect the environment and reduces the time spent collecting firewood. The stoves are made by local people and use local materials.
Projects such as this one in Nepal help to create employment in rural areas.

Q Can you explain what is meant by small-scale, local development?

PRACTICE

1 Using named examples, explain what is meant by large-scale development.

2 With reference to a named example, explain the advantages of small-scale development in LEDCs.

Sustainable development is a very important geographical concept. Make sure that you refer to it wherever possible in your exam answers.

Resources and energy

THE BARE BONES
➤ Natural resources are used to generate energy for homes and industry.
➤ They can be classified as renewable and non-renewable.
➤ Extracting and using natural resources has an impact on people and the environment.

A Resources

KEY FACTS

1 A resource is a supply of something that people can use.

2 All industries rely on an input of resources. As a country develops and begins to <u>industrialise</u>, it <u>uses more resources</u>.

3 Resources can be **natural** and **human**:

- **Natural resources** come from the **earth** (e.g. rocks or minerals). The **landscape** itself is a resource. **Attractive** and **interesting** landscapes are important resources for the **tourist industry**.

- **People** provide **human resources**. For example, a factory **steelworker** has **important skills** that are a **resource** to the steel company.

4 Natural resources can be **classified** (organised) into two groups:

Non-renewable Resources exist in limited supplies and cannot be replaced once they have been used. Non-renewable resources include coal, oil and natural gas.

Renewable Resources will not run out and can be used again and again. Renewable resources include wind, sun and water.

Q Explain the difference between renewable and non-renewable energy sources.

B Extracting natural resources

1 Natural resources such as **coal** and **oil** have to be extracted from the earth.

2 Coal and oil are called **fossil fuels**. Fossil fuels are formed through the fossilised remains of dead plants and animals.

3 Extracting fossil fuels can have an impact on people and the natural environment.

Case Study – Coal

Coal is a sedimentary rock. It can be used for generating energy, heating homes and transport. Coal forms in layers, which are called seams.
Coal-mining involves extracting coal from within the earth. In the past, coal-mining involved removing coal from seams close to or at the earth's surface. As technology has developed, it has become possible to mine deeper beneath the earth's surface. Accessible supplies of coal in the UK are now beginning to run out. As supplies of coal become more difficult to mine, the cost of mining increases.
The UK now imports some of its coal supplies from other countries, including LEDCs. Imports can be cheaper and many LEDCs have very good reserves of coal.

Remember
As LEDCs begin to industrialise, the global demand for energy will rise. More pressure will be placed on limited supplies of natural resources.

B

Creates employment and income for people and places.

Benefits of coal-mining – mainly social or economic

Strong communities can develop in mining towns, for example in the Welsh Valleys.

Coal is used to generate power for factories and industry. Industries generate wealth and income for a country and its workers.

When a mine is opened, other services and industries are attracted to locate close by. This sets off the multiplier effect (see Development, p.122).

Mining can be difficult and dangerous.

Problems created by coal mining - mainly environmental

Large amounts of land are destroyed for opencast mining. It is hard to re-landscape old mining areas.

Once a coalfield is exhausted and all the coal has been mined, it will close. This can have serious impacts on mining communities.

Mining creates heavy traffic and noise pollution.

Coal-mining produces waste rocks and materials. These materials are stored in piles called spoil tips. Spoil tips create unattractive landscapes.

Q Can you explain what a fossil fuel is?

C Generating energy

1 Natural resources are used to produce energy, usually in the form of **electricity or gas**. This **energy** is used to **power homes, transport and industry**. **Most** of the UK's energy is **produced** in **power stations**.

Non-renewable energy sources	Renewable energy sources
Coal, oil, natural gas, fuelwood (mainly used in LEDCs), nuclear power	Hydro-electric power, wind power, tidal power, solar power, geothermal energy (using the earth's natural heat)

2 The **demand** for energy is **increasing** both in the UK and at a **global level**. The global demand for energy is growing as LEDCs develop and use more energy to power homes and industries.

3 There are important **differences** in **patterns** of **energy demand** across the world:

Most of the world's population lives in LEDCs (80 per cent of the world's population), yet LEDCs only consume (use) 20 per cent of the world's energy.

Only 20 per cent of the world's population lives in MEDCs, yet MEDCs consume 80 per cent of the world's energy.

4 Supplies of **natural resources** are **running out** in many MEDCs. They have begun to import from LEDCs and to develop **alternative energy** sources, such as **solar power**, which tend to be sustainable and **environmentally friendly**.

Q Can you explain why the global demand for energy is rising?

PRACTICE

1 Give two uses for coal as a natural resource.

2 Describe the impacts of coal-mining on people and the environment.

Non-renewable energy

THE BARE BONES

➤ Most of the world's energy is generated by burning fossil fuels.
➤ Burning fossil fuels can cause many environmental problems.
➤ Acid rain can be caused by pollution from power stations.
➤ Nuclear energy is used as an alternative to fossil fuels in some countries.

A Fossil fuels

KEY FACT

1 Oil, coal, natural gas and fuelwood are fossil fuels.

2 Burning fossil fuels, particularly coal, generates **most** of the **world's energy supplies**.

3 Fossil fuels account for over **75 per cent** of the **UK's energy supplies**, as in most MEDCs.

4 Burning **fuelwood** provides the **main source of energy** in many LEDCs.

5 Fossil fuels are **non-renewable** and there are **finite** (restricted) **supplies** of them. Unless **alternative** energy sources can be developed on a large scale, the world could face an **energy crisis** in the future.

6 Burning fossil fuels has a serious **impact** on the **environment**. Many enviromental groups and international organisations are trying to reduce their use.

	✓ Advantages	✗ Disadvantages
Coal	• Available in many countries. • Large reserves remain relatively untouched.	• Burning coal creates pollution and is linked to acid rain. • Releasing carbon dioxide is linked to the onset of global warming (see Climate, p.58). • Mining can be dangerous and damages the environment.
Oil	• An efficient source of energy. • A diverse energy source that can be used in cars, engines and power stations. • Easily transported via pipelines, tankers and lorries.	• Air pollution is created. • Releases carbon dioxide. • Danger of oil spills in seas and oceans.
Natural Gas	• An efficient source of energy. Little waste is produced. • Cleaner than other fossil fuels (produces less pollution). • Easy to transport via pipelines.	• Some air pollution is created. • Releases carbon dioxide. • Highly flammable (fires and explosions can be caused).
Fuelwood	• Low cost, often free. • Widely available. • Can become a renewable energy source if new trees are planted.	• Releases carbon dioxide. • Supplies are running low in many rural areas of LEDCs. • Clearing trees can cause soil erosion and desertification (the spread of desert conditions).

Q Can you list the advantages and disadvantages of using fossil fuels to generate energy?

B Acid rain

1 The burning of fossil fuels can create <u>environmental pollution</u>.

2 A serious environmental problem created by this pollution is **acid rain**.

Remember
Acid rain is not the same as global warming or the greenhouse effect. Make sure you understand what each of these terms mean. They are all different issues.

Acid rain is rain that has a higher than average level of acidity. Natural rainfall has an acidity level (pH) of about 5.6, whereas acid rain has a pH of about 4.5. The lower the pH, the higher the level of acidity.
Acid rain is caused when gases such as sulphur dioxide (SO) and nitrogen oxide (NO) are released into the atmosphere as pollution from power stations and factories. Sunlight converts these gases into sulphuric acid and nitric acid. These acids dissolve in moisture in the atmosphere, turning them into a weak acid. Acids then fall to the earth as rainfall and enter the hydrological cycle.

3 Acid rain can be carried large distances by the **wind**. This means that **damage** can be caused **far away from the source** of the pollution.

4 Large areas of **Scandinavia**, including Norway and Sweden, have been affected by acid rain, possibly caused by pollution from the UK and southern Europe.

5 Acid rain can kill fish and other animals in lakes and rivers. In Sweden, 20 per cent of **lakes** are seriously damaged by acid rain. There is also evidence of high acidity in lakes and streams in Wales and the Lake District.

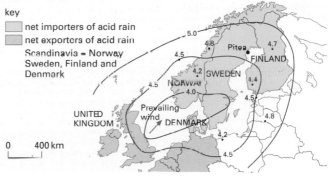

key
- net importers of acid rain
- net exporters of acid rain
Scandinavia = Norway
Sweden, Finland and
Denmark

6 Acid rain can damage and even kill **trees** in forest areas, by falling directly onto the leaves or by entering the soil and attacking the roots.

Q Can you list three impacts of acid rain?

7 Acid rain may affect farming by reducing **crop levels**.

C Nuclear energy

1 As supplies of fossil fuels run low and environmental concerns grow, many countries have developed <u>nuclear power</u>.

Q Describe the advantages and disadvantages of using nuclear energy.

2 Nuclear power is **cleaner** and more **efficient** than fossil fuels, and does not release greenhouse gases or create acid rain. However, many people are concerned about the **safety** of nuclear power, particularly after the accident at **Chernobyl** in the Ukraine, in 1986. The nuclear debate is a controversial one, with strong opinions and arguments on both sides.

PRACTICE

1 List two disadvantages of using fossil fuels to generate energy.

2 Explain why acid rain is an international problem.

Renewable energy

THE BARE BONES
➤ Many countries are now beginning to use sources of renewable energy.
➤ Renewable energy sources will not run out and tend to produce little pollution.
➤ Using renewable energy sources could help the UK and other countries to meet the targets set by the Kyoto Agreement.

A Renewable energy

KEY FACT ▶

1 In recent years many countries have begun to develop renewable energy sources.

- **Renewable energy sources** are not going to run out and can be used again and again.

- Renewable energy sources tend to be **clean** and **environmentally friendly;** they produce little **waste** or **pollution**.

2 Using renewable energy sources means that energy generation can be sustainable.

3 Renewable energy is sometimes called **alternative energy**. This is because renewable energy provides people with an alternative to fossil fuels and other non-renewable energy sources.

4 Alternative energy sources have been developed for two main reasons:

> International organisations and environmental groups are very concerned about the effects of burning fossil fuels on the environment.

> People are concerned about whether there will be enough natural resources to meet future energy demands.

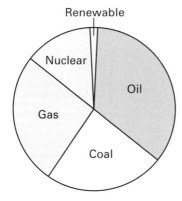

Energy mix for UK

Remember
Using renewable energy sources is more sustainable than using non-renewable ones, as they can be used over and over again, with little impact on the environment.

Q Can you list the benefits of using renewable energy sources?

B The Kyoto Agreement

1 The **Kyoto Treaty** was drawn up in **Kyoto, Japan** in **1997**. It is an **international agreement** that aims to **reduce emissions** of **carbon dioxide** into the atmosphere and therefore help to protect the environment.

2 The **USA** has **pulled out** of the agreement, which has caused a lot of debate over whether the **targets set** can be met. The USA is the world's **greatest consumer of energy** and therefore **creates** a lot of **pollution**.

B

3 **Developing renewable energy sources** will play an important part in helping to **meet** the **targets** set by the Kyoto Treaty, however, there are limitations to using renewable energy:

- The **cost** of **researching** and **developing** renewable energy sources is **high**.

- Many renewable energy sources do not generate large enough supplies of energy to completely replace fossil fuels.

- Renewable energy sources are **mainly used** in MEDCs. Most **LEDCs do not** have the **money** or resources to develop renewable energy sources.

Q Make a list of the problems and limitations of using renewable energy.

C Generating renewable energy

Like non-renewable energy sources, using renewable energy can have **advantages** and **disadvantages**.

Remember
The cost of setting up a renewable energy project is often very high. However, once it has been set up, the cost of running the project is generally low.

	✓ **Advantages**	✗ **Disadvantages**
Wind power	• Wind turns blades on large turbines to generate electricity. • Wind power is a very clean source of energy; no waste or pollution is produced. • Relatively cheap to produce. • Burning coal creates pollution and is linked to acid rain.	• Wind turbines need to be placed on exposed, undeveloped landscapes. This usually means areas of environmental importance. • Lots of turbines are needed to produce a useful amount of energy. • Many people feel that wind turbines are unattractive and noisy.
Solar power	• Solar panels or photo-voltare cells use sunlight to produce electricity. • Solar power is a very clean source of energy; no waste or pollution is produced. A cheap and efficient source of energy.	• The technology needed to produce solar power is expensive to develop and install. • Solar energy cannot be generated at night or in very cloudy conditions.
Tidal power	• Moving water in tidal areas turns turbines to generate electricity. • Tidal power is a clean energy source. It is relatively cheap to produce. • Tidal power can produce a large amount of energy.	• The technology needed to produce tidal power is expensive to build and install. • Producing tidal power can affect coastal ecosystems. • There are few sites available for tidal power.
Geothermal power	• Water is heated using the earth's natural heat. This produces steam, which turns turbines. • There are many potential sites for geothermal energy. • Many potential sites are in LEDCs.	• Sulphuric gases are produced as a by-product of the geothermal energy. • It is expensive to develop geothermal energy and maintenance costs can be high.

Q What are the advantages and disadvantages of using wind power?

PRACTICE

1 Give two reasons why many countries are now developing renewable energy sources.

2 a) Using the pie chart opposite, what's the main source of energy used in the UK?

b) How may the energy mix of the UK change in the future?

3 For a named renewable energy source that you have studied, explain the advantages and disadvantages of using that energy source.

This section provides two things:

1. General exam advice, including how to prepare and how to tackle different kinds of questions and papers.

2. Some sample questions with answers and examiner's explanations (in yellow boxes), showing why these answers achieve full marks.

About GCSE Geography exams

1. Foundation Tier or Higher Tier?

In GCSE exams, you can be entered at different levels: Foundation Tier or Higher Tier. This is to give everyone the chance to get the best grade they can. Your teacher will have decided which is the best level for you. Questions on Foundation Tier papers are very clear and easy to understand. You usually write your answers in an answer booklet and the space given shows you about how long your answer should be.

Higher Tier papers often include more complex resources to test your skills of interpretation and analysis. The questions are usually more demanding and they may be 'open-ended', which means that there could be a range of possible answers. In these types of questions you can use a variety of examples and suggest different reasons for your answers.

2. Understanding the questions

GCSE examinations are not trying to catch you out. Questions may not be about places you have studied, but they will give you the opportunity to show what you know, understand and can do in geography.

It's really important, when you are keen to get on with an exam paper, that you read and understand the question, or you may answer it in the wrong way and lose marks. Here are some words commonly used in geography exams. Do you know exactly what they are telling you to do?

Annotate – Add notes or comments to labels on maps or diagrams to explain what they show.

Compare – Identify and write down the similarities and differences between features or places stated in the question.

Complete – You might be asked to add the remaining parts of a diagram, map or graph.

Contrast – Write down and point out clearly the differences between the features or places.

Define – Write a definition of, meaning describe accurately or explain the meaning of.

Describe – Write down details about what is shown in a resource such as a map or diagram.

Discuss – Usually requires a longer answer, describing and giving reasons or explaining the arguments for and against.

Draw – You might be asked to draw a sketch map or diagram with labels to identify particular features.

Explain (or Account for) – Give reasons for the location or appearance of a particular feature.

Factors – Reasons for something such as the location of particular geographical features.

Give your views – You might be asked to say what you think or what another person or group might think.

Identify – Name, locate, recognise or select a particular feature or features (usually from a map, photograph or diagram).

Locate – Write down where a feature or place is.

Mark – You may be asked to indicate or show where particular features are on a diagram or map.

Name, state or list – Write down accurate details or features.

Study – Look carefully at a resource and think about what it shows.

With reference to (or refer to) examples that you have studied – You need to include details about specific case studies or examples when explaining the reasons for a particular answer.

With the help of or using the information provided – Be sure to include examples from the information on the paper to explain your answer.

3. Decision-making or problem-solving papers

Some geography syllabuses include a different kind of exam paper, about decision-making, problem-solving or evaluating an issue. This type of exam is not discussed further in the book because it involves working directly with many resources, but here are some tips on how to deal with it.

For this kind of exam you are given a number of different resources with which to investigate a geographical or environmental issue and it tests your skills in using them. Some exam boards send a booklet of the paper's resources to the school a few weeks before the exam so you can use it to prepare.

Exam boards usually tell your teachers which part of the geography syllabus the issue will come from, so they can cover relevant ideas and topics to help you prepare. Although the structure of questions varies between boards, they all follow a similar sequence.

Early questions ask you to interpret resources and outline the issues involved. You are then asked to evaluate or suggest different solutions to the problems. Finally, you are asked to explain which solution you think is best.

4. Preparing for problem-solving papers

- Look at past papers or specimen papers.

- Read through the resources and familiarise yourself with them.

- Make lists of the types of information they show and the geographical ideas and processes that are relevant to the issue. Make sure you understand the geographical terms used. Improve your background knowledge of the issue.

- Practise drawing sketch maps you could use in the exam.

5. Practise answering GCSE questions

Try including one or two practice questions in your geography revision each week so that you practise using what you are revising. Try doing this with a friend too, then check your answers. Explain to each other why you included particular facts or examples.

Remember to use real examples about real people and real places. Don't be vague, name examples, give facts and figures and use the correct geographical terms.

6. In the exam

- Read the instructions carefully.

- Choose which questions to answer. Read the whole paper and decide which questions you can answer all the parts of. Think about which case studies you could use.

- Plan the longer answers. Write brief notes to remind yourself what to include and follow this plan as you write. Don't spend too long on plans, though, as you'll need the time for writing.

- Answer the question – only write what the question is asking for.

- Unless the instructions say otherwise, plan to spend the same amount of time on each question. You can work out how long this should be before you start.

- A small number of marks are given for spelling and punctuation. It is especially important to make sure you learn how to spell geographical terms.

Interpreting maps and aerial photographs – explaining location

You may get questions where you're given a photograph alongside a sketch, such as the example here (see sketch and photograph opposite). Details in the photograph may not be clear. You must use your knowledge and understanding to explain your answers.

The Merry Hill Shopping Centre, Dudley, West Midlands, is one of the biggest shopping centres in the UK. Look at the photograph and outline sketch and use the sketch to answer the questions.

Notice the difference between the Foundation and Higher Tier questions!

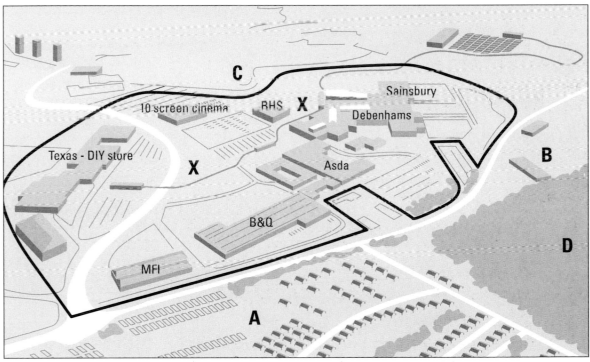

Foundation

a) Find letters **A**, **B**, **C** and **D** on the sketch. Match each letter with one of the labels: factories, canal, housing, woodland.

   A. *Housing* **B.** *Factories* **C.** *Canal* **D.** *Woodland*

b) **X** on the sketch is a monorail. Why do you think it was built ?

   Because of the size of the shopping centre and the distance between different stores. The monorail makes it easier to travel around the centre. It connects the car-parking areas to the major store locations in different parts of the centre.

Higher

a) On the outline sketch of Merry Hill, add labels to describe the areas marked **A** and **B**.

Notes on type and layout of housing and factories.

2 marks

b) Describe the layout of car parking at Merry Hill.

Describe where car parks are in relation to access points, shopping areas and cinema.

2 marks

Foundation

c) Suggest why this place was chosen for a large shopping centre. Use these words:

Transport   Site   Workers   Customers

Large amount of land needed, land price, accessibility, transport links, competition, attitude of local planners.

4 marks

Higher

c) What do the developers of shopping centres need to consider when choosing a location for a shopping centre such as Merry Hill?

4 marks

Large amounts of land are needed for shopping centres such as Merry Hill because of their size and the large area needed for parking and big shops or stores. The cost of this land is important. Larger amounts of cheaper land are more likely to be found on the edge of towns and cities.

Accessibility is important as these centres need large numbers of customers. Therefore, developers look for sites that have good transport links and are near to a number of large towns or cities.

The attitude of local planners to large out-of-town shopping centres and whether there are other similar centres nearby can also be important factors.

This answer would gain all 4 marks, for giving at least 4 well-explained reasons and showing good understanding of the factors influencing the location of large out-of-town shopping centres such as Merry Hill.

Foundation

d) i) Write down two advantages of a shopping centre such as Merry Hill.

ii) Write down two disadvantages of a shopping centre such as Merry Hill.

4 marks

i) Shopping centres such as Merry Hill have a large number and range of shops. They have several of the main stores, such as Sainsburys, Asda and Debenhams (major chain stores). They also have large amounts of free parking and often have other attractions, such as a cinema or leisure complex.

ii) They take up large areas of land on the edge of towns and cities and can add to traffic congestion on roads in these areas. They are not easy to get to for people without cars because they are often not well served by public transport. The success of large out-of-town shopping centres can lead to shops closing down in local neighbourhood shopping centres and in town and city centres. People without cars can suffer if this happens.

These answers score full marks because each point is fully explained. As a general rule, one mark is awarded for each well-explained point in an answer!

Ordnance Survey maps in GCSE Geography examinations

Most GCSE Geography examinations include at least one question based on an Ordnance Survey map extract. The OS map on page 136 shows the area around Newport in South Wales. A map like this could be used in questions about industrial location, transport routes or urban settlements. Here are some typical questions that you could be asked:

1. Give the 4-figure grid references for the grid squares where the following are to be found: a) Llanwern Steel Works; b) Newport Docks; and c) the power station at the mouth of the River Usk.

   a) 3586, 3686, 3786, 3886, 3986; b) 3184, 3284, 3185 and the corner of 3285; c) 3283

2. Work out the approximate area covered by the Llanwern Steel Works.

   Approximately 5 square km

3. Use evidence from the map to describe where you think Newport's CBD (Central Business District) is located on the map.

> Use a ruler and the scale to work out area covered. On a 1:50 000 map, 2cm = 1km. The steel works cover approx. 5 grid squares, with each grid square being one square kilometre.

> Think about buildings and land uses you would expect to find in a CBD. Find the map symbols for these and describe their location on the map using grid references.

- Look for the main public buildings – cathedral, historic buildings, bus station, etc.

- The CBD is in parts of grid squares 3187, 3087, 3088 and 3188.

Remember!

- Use 6-figure grid references to describe the exact locations of features.

- Use 4-figure grid references to describe the location of grid squares if you are locating a larger area.

- You might be asked to describe and explain (give reasons for) the location or distribution of particular land uses on an OS map. A sketch map can help you to do this:

4. Why do you think that the Llanwern Steel Works were built here?

   The coastal site is good for a large steel works because there are large amounts of flat land and water needed for cooling in the steel making process. Steel works also need a large supply of power and this is provided by the power station at Uskmouth. Raw materials can be imported in large quantities through the docks nearby at Newport (local supplies of coal were also available from coal mines in the South Wales coalfield). The Llanwern site is also well served by good transport links (road and rail).

> Sometimes, drawing an annotated (labelled) sketch map is a quick and effective way of picking up marks!
>
> Quickly draw an approximate outline of the main features of the area (in this case, the coastline, river and estuary). Mark a few of the main transport routes (M4, A-roads, railway). Shade the main settlements in one colour and industrial land use in another colour.

> Always try to use appropriate geographical language when you are writing answers. For example, if you are describing features on a map, use compass directions to describe where they are in relation to each other (e.g. 'the river flows from the south west to the north east'). Also use the scale of the map to help you describe how far apart features are (e.g. 'the power station is located 2km to the south west of the steel works').

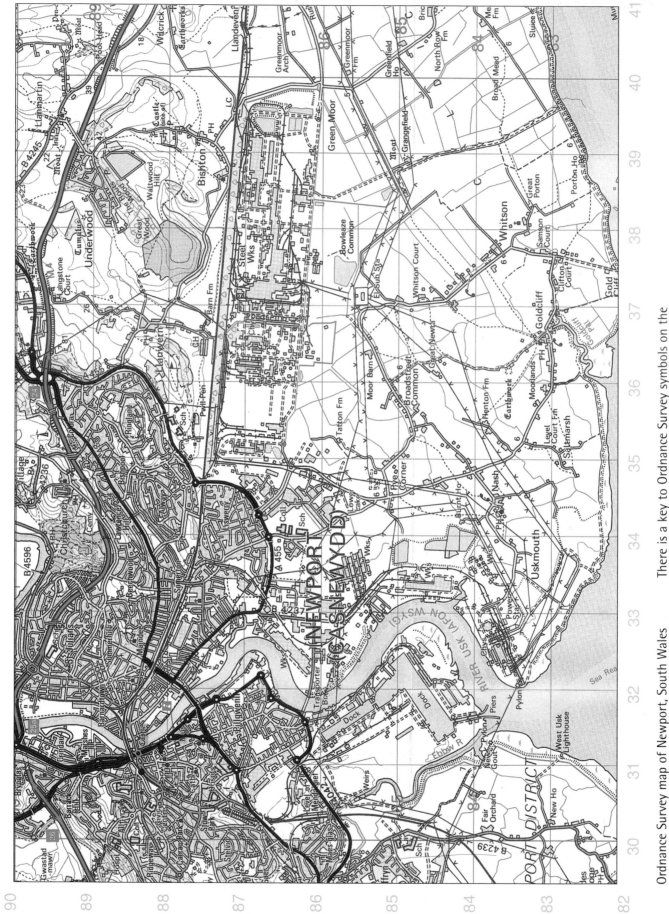

Ordnance Survey map of Newport, South Wales

See the sample questions on page 135.

There is a key to Ordnance Survey symbols on the page opposite.

Key to Ordnance Survey symbols

ROADS AND PATHS
Not necessarily rights of way

Service area M 6   Elevated   Motorway (dual carriageway)
Junction number 41

Motorway under construction

Unfenced   Footbridge   Trunk road
A 6 (T)

Dual carriageway   Main road
A 592

Main road under construction

Secondary road
B 5305

Narrow road with passing places
A 855   B 885

Bridge   Road generally more than 4 m wide

Road generally less than 4 m wide

Other road, drive or track

Path

Gradient : 1 in 5 and steeper
1 in 7 to 1 in 5

Gates   Road tunnel

Ferry P   Ferry V   Ferry (passenger)   Ferry (vehicle)

PUBLIC RIGHTS OF WAY
(Not applicable to Scotland)

.... Footpath

- - - - - - Bridleway

-·-·-·-· Road used as a public path

-+-+-+-+- Byway open to all traffic

ANTIQUITIES

VILLA   Roman   ⚔ Battlefield (with date)   + Position of antiquity which cannot be drawn to scale
Castle   Non-Roman   ☆ Tumulus

𝔪   Ancient Monuments and Historic Buildings in the care of the Secretaries of State for the Environment, for Scotland and for Wales and that are open to the public

The revision date of archaeological information varies over the sheet

RAILWAYS

——— Track multiple or single   ┼┼┼┼ Freight line, siding or tramway
—‖—‖— Track narrow gauge   ●■○ Station   (a) principal (b) closed to passengers
—‖‖— Bridges, Footbridge   ‖LC Level crossing
—▥┅┅▥— Tunnel   ▥▥▥ Embankment
—✕— Viaduct   ▤▤ Cutting

WATER FEATURES

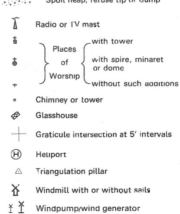

Marsh or salting
Towpath   Lock
Aqueduct
Weir   Canal   Ford
Lake   Normal tidal limit
Bridge
Footbridge
Canal (dry)
Slopes   Cliff
Flat rock
Sand   Dunes
Mud
High water mark
Low water mark
Lighthouse (in use)
Beacon
Lighthouse (disused)
Shingle

TOURIST INFORMATION

ℹ Information centre, all year / seasonal   ▲ Youth hostel
Viewpoint   Selected places of tourist interest
P Parking   ✆ ✆ Telephone, public/motoring organisation
✕ Picnic site   ⌐ Golf course or links
⚊ Camp site   PC Public convenience (in rural areas)
🚐 Caravan site

ABBREVIATIONS

P	Post office	CH	Clubhouse
PH	Public house	PC	Public convenience (in rural areas)
MS	Milestone	TH	Town Hall, Guildhall or equivalent
MP	Milepost	CG	Coastguard

GENERAL FEATURES

⋏——⋏——⋏ Electricity transmission line (with pylons spaced conventionally)
> - -> - -> Pipe line (arrow indicates direction of flow)
ruin   Buildings
■ Public buildings (selected)
Bus or coach station
Coniferous wood
Non-coniferous wood
Mixed wood
Orchard
Park or ornamental grounds
Quarry
Spoil heap, refuse tip or dump
Ⲧ Radio or TV mast
Places of Worship — with tower / with spire, minaret or dome / without such additions
∘ Chimney or tower
⌗ Glasshouse
+ Graticule intersection at 5' intervals
Ⓗ Heliport
△ Triangulation pillar
Windmill with or without sails
Windpump/wind generator

HEIGHTS

——50—— Contours are at 10 metres vertical interval
·144 Heights are to the nearest metre above mean sea level

BOUNDARIES

—·—·— National
—·—·— London Borough
National Park or Forest Park
[NT] National Trust
NT always open

ROCK FEATURES

outcrop   cliff   600   scree

Heights shown close to a triangulation pillar refer to the station height at ground level and not necessarily to the summit.

Ordnance Survey maps in questions about national parks

Look at the Ordnance Survey map of Keswick opposite. Here are some typical questions that you might be asked about it. The first parts of a GCSE question usually test your map skills. Examples might be:

1. a) Why might a farmer in grid square 2519 want to open a caravan site there?

   | 2 marks |

   Fairly flat land; near to roads and passing traffic; good views.

   b) Give an example of one person or group of people who might object to this idea and say why they might object.

   | 2 marks |

   Other farmers with caravan sites nearby might fear the competition; nearby residents, or walkers, who enjoy the farmer's land as it is.

   c) Give three reasons why the area in grid square 2618 might attract many visitors and become a 'honeypot site' (see p.111).

   | 3 marks |

   Beautiful view over lake; car parking; start of some interesting walks.

Later parts of a question test your understanding and application of what you have learnt in a topic to an OS map. Here are some examples:

2. Weekend traffic, except for local residents, could be banned from some roads in the Lake District. If this was proposed for the minor road between Borrowdale Gates (2518) and Hawes End (2521):

   a) Name one group of people who might be in favour of the proposal and say why.

   | 3 marks |

   Walkers/hikers: greater peace, less traffic noise; farmers who might open tearooms if the route becomes popular.

   b) Name one group of people who might be against the proposal and say why.

   | 2 marks |

   Motorists who enjoy the small, scenic roads; farmers who might lose income from bed and breakfast /tearooms if cars no longer pass their door.

Towards the end of a question you are often asked to write in more detail and to use your own knowledge rather than information given on the map.

3. Areas such as that shown on the OS map extract attract large numbers of visitors. Give your views on how the damage they might cause to the natural environment could be limited.

   | 5 marks |

> You should write about damaged areas in general, but refer to one or more case studies to give specific information to back up your points. You could describe the impacts of visitors on the environment at 'honeypot' sites and explain how particular strategies could help to reduce these impacts.

Ordnance Survey map of the Keswick area
See the sample questions opposite.

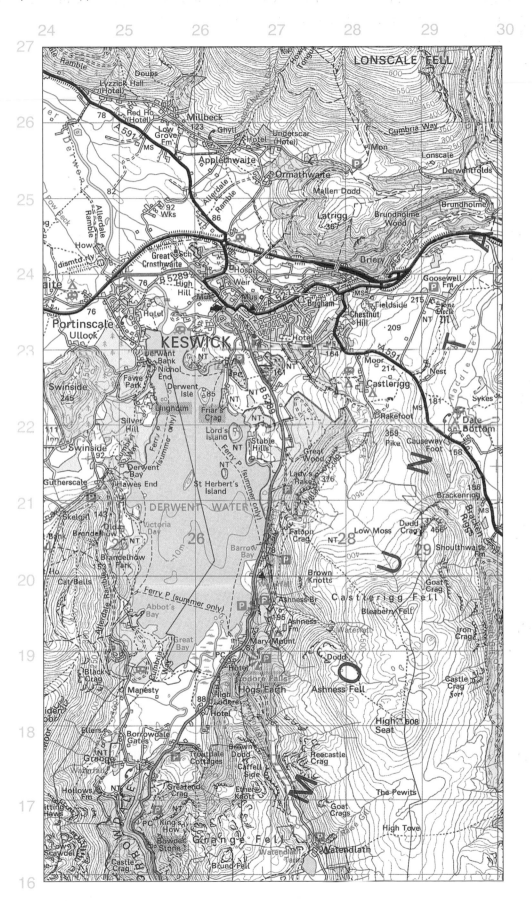

Cartoons in examination questions

Cartoons are sometimes used in questions where the examiner wants to find out what you know and understand about the values and attitudes that particular groups of people might have. In the examples shown here, the reason for including the cartoons is to focus on what attracts people to move to cities (pull factors) and the reality of what they find when they get there. Here are some questions about the cartoons opposite that you could try.

Foundation Tier

Look at the two cartoons, b) and d), on the page opposite, which comment on features of urban growth in Less Economically Developed Countries.

Choose **ONE** of the cartoons and explain the situation shown.

2 marks

Higher Tier

5 marks

On the page opposite there is a selection of cartoons about life in large cities. Use the cartoons to help you explain why the attitudes of migrants to cities may change once they get there.

The first part of each cartoon strip opposite shows a pull factor that attracts people to move to a city. The second part shows the reality of what they often find. For example, the first strip suggests that one of the reasons why people move to large cities in LEDCs is the prospect of better paid jobs compared with the low wages and poverty they may experience in the countryside. However, what they often find is that life in the cities may not be much better. Here are some reasons why:

- *There may be a lot of unemployment because there are so many migrants looking for jobs.*

- *Basic needs such as food, fuel and water cost more because there is a great demand for them.*

- *Most of the food and fuel has to be brought into the areas where the migrants live from outside the city, which also makes it more expensive.*

When answering questions on cartoons, remember that one problem with them is that they can give a fairly narrow and stereotypical view. For example, for many of the migrants the quality of life in the squatter settlements and slums may genuinely be better than in the rural areas they left. The migrants may actually be able to improve their quality of life in the longer term, so the effect shown in the cartoon may not apply to everyone.

The examiner will be hoping that you will show what you understand about these factors. Using some information about places that you have studied will help you to get all of the marks that are available.

Topic checker

- Go through these questions after you've revised a group of topics, putting a tick if you know the answer, a cross if you don't – you can check your answers at the bottom of each page.
- Try these questions again the next time you revise . . . until you've got a column that's all ticks! Then you'll know you can be confident.

Plate tectonics

1	What causes the plates that make up the earth's crust to move?	☐	☐	☐
2	Why are most of the world's volcanoes and earthquakes found near to plate boundaries?	☐	☐	☐
3	What happens at a constructive plate boundary?	☐	☐	☐
4	At what type of plate boundary do volcanoes normally form?	☐	☐	☐
5	What point on the earth's surface is likely to suffer the greatest damage from an earthquake?	☐	☐	☐
6	What is the difference between magma and lava?	☐	☐	☐
7	Why do earthquakes sometimes occur in the same place as volcanoes?	☐	☐	☐
8	What is the difference between the primary and secondary effects of an earthquake?	☐	☐	☐
9	What factors influence the amount of damage caused by earthquakes?	☐	☐	☐
10	What can be done to restrict the damage caused by earthquakes?	☐	☐	☐

Answers 1 Convection currents 2 The earth's crust is weaker and there is greater tectonic activity near plate boundaries 3 Two plates move apart and magma rises to form new crust, e.g Mid-Atlantic Ridge 4 Destructive and constructive plate boundaries 5 Near to the epicentre of an earthquake 6 Magma is molten rock beneath the surface of the earth; Lava is molten rock that has flowed out over the surface of the earth 7 Instability near plate boundaries and the huge pressure building up in subduction zones where volcanoes form can trigger earthquakes 8 Primary – immediate impacts (e.g roads and buildings collapsing); Secondary – result from damage caused by the initial tremors, e.g fires caused by gas leaks from pipes fractured by the earthquake 9 Strength of the earthquake; density of population and buildings in areas affected; distance from epicentre; type of construction and ground on which they are built; time of day when the earthquake occurs 10 Restrict building in areas prone to earthquakes; build earthquake-proof buildings that 'sway' rather than collapse; education and communication about what to do in an earthquake; earthquake monitoring and warning systems

Rocks and weathering

11 How were igneous rocks formed?	☐ ☐ ☐
12 How were metamorphic rocks formed?	☐ ☐ ☐
13 What is the difference between weathering and erosion?	☐ ☐ ☐
14 What is the difference between physical and chemical weathering?	☐ ☐ ☐
15 Why are there so few streams in limestone areas?	☐ ☐ ☐

Glaciation

16 What is a corrie?	☐ ☐ ☐
17 What is the difference between abrasion and plucking?	☐ ☐ ☐
18 What different types of moraine are there?	☐ ☐ ☐
19 How can moraines provide evidence about the movement of a glacier?	☐ ☐ ☐
20 Describe two features of glacial deposition.	☐ ☐ ☐

The river system

21 What is the difference between throughflow and groundwater?	☐ ☐ ☐
22 Which processes transfer water in a drainage basin?	☐ ☐ ☐
23 What is the discharge of a river?	☐ ☐ ☐
24 How can flood hydrographs help to predict where and when there are greater flood risks?	☐ ☐ ☐
25 Why will building more houses on a flood plain increase the risk of flooding?	☐ ☐ ☐

11 When magma from inside the earth cooled and solidified 12 When intense heat and pressure changed existing rocks (e.g limestone into marble) 13 Weathering is the breaking down of rocks either at the surface of the earth or underneath soil without any movement of these rocks; Erosion is the wearing away of rocks by water, ice or wind (i.e involving movement) 14 Physical weathering usually results from large changes in temperature or pressure; Chemical weathering occurs when chemicals dissolved in water attack and break down rock surfaces 15 Limestone is a permeable rock (allowing water to pass through), so there are few streams on the surface 16 Glaciers begin in hollows on the colder side of mountains. As the hollow becomes bigger, a corrie is formed 17 Abrasion – rock fragments and ice act like rough sandpaper wearing away rocks over which ice moves; Plucking occurs when meltwater under a glacier freezes onto rock surfaces, pulling away large bits of rock when the glacier moves forward 18 Medial, lateral and terminal 19 Different types of moraine are deposited at different places and at different stages in the movement of the glacier (e.g. terminal moraine will show the furthest point reached by the glacier) 20 Drumlins: mounds of boulder clay that have been shaped by the ice. Ice moves over them to form egg-shaped hills. Erractics: rocks transported many miles by the glacier and found in an area of different type 21 Throughflow is water flowing beneath the surface but above the water table; Groundwater is water below the water table 22 Evaporation, transpiration, precipitation, infiltration, surface run-off (overland flow), throughflow, groundwater flow 23 The amount of water flowing in a river past a particular measuring point over a given period of time (measured in cumecs) 24 They show the relationship between rainfall and the discharge of a river in different weather conditions The flood risk is greater when the discharge of a river has a short lag-time and rises rapidly after a period of rainfall 25 Water reaches rivers more rapidly through drains and by flowing off impermeable surfaces on and around buildings

26 Why can the planting of trees on valley slopes help to reduce the risk of flooding?	☐	☐	☐
27 Name the four different methods by which material can be transported by a river?	☐	☐	☐
28 What is the difference between abrasion and attrition?	☐	☐	☐
29 Why do deltas sometimes form at the mouths of very large rivers?	☐	☐	☐
30 What is eutrophication and how can it affect plants and wildlife in a river?	☐	☐	☐

The sea at work

31 What is the difference between the swash and the backwash of a wave?	☐	☐	☐
32 How does the fetch of a wave affect its erosional power?	☐	☐	☐
33 How are beaches formed by destructive waves likely to be different from those formed by constructive waves?	☐	☐	☐
34 What is the name given to the process of wave erosion resulting from changes in pressure caused when waves trap air in cracks in rocks and then retreat?	☐	☐	☐
35 Why might the building of groynes in one location result in an increase in erosion further along a coastline?	☐	☐	☐
36 How do stacks form?	☐	☐	☐
37 What is the difference between 'hard' and 'soft' approaches to coastal management?	☐	☐	☐
38 What are gabions and how do they help to reduce coastal erosion?	☐	☐	☐
39 What are the disadvantages of building sea walls to protect areas of coastline?	☐	☐	☐
40 What are the advantages of using 'beach nourishment' to protect a coastline?	☐	☐	☐

26 Increases interception and slows down the rate at which water reaches the ground and into rivers 27 Traction, saltation, suspension and solution 28 Abrasion – material carried by rivers scrapes away the bed and banks; Attrition – material carried by rivers breaks down into smaller fragments as it knocks into other material 29 Very large rivers carry vast amounts of sediment; As these rivers flow into the sea, they slow down, depositing material faster than it can be removed by the sea 30 An increase in nitrate levels in rivers caused by pollution, encouraging the growth of plants, particularly algae; Algae uses up oxygen and blocks out light affecting wildlife and turning water green 31 Swash – movement of water up a beach; Backwash – movement of water back down a beach 32 Waves with a larger fetch have more energy and stronger erosional power 33 Constructive waves have a stronger swash and build up beaches;

Destructive waves have a stronger backwash and erode material from a beach 34 Hydraulic action
35 Groynes prevent sediment from moving along a coastline, depriving other areas of sediment; Beaches further along the coast will be smaller and more open to erosion by waves 36 A stack forms when the roof of a natural arch in a headland collapses 37 Hard approaches try to protect coastlines by deflecting or breaking up wave energy (working against natural processes); Soft approaches work with natural processes to build up natural defences 38 Gabions are wire baskets filled with rocks that try to reduce erosion by breaking down wave energy as waves bass between the rocks 39 Sea walls protect the most vulnerable land uses by deflecting strong waves 40 Beach nourishment builds up beaches to break up incoming waves – it works with natural processes and looks more natural so has less visual impact

Weather and climate

41 What is the difference between weather and climate?	☐	☐	☐
42 What is the difference between a continental climate and a maritime climate?	☐	☐	☐
43 How does the weather associated with a depression differ from the weather associated with an anticyclone?	☐	☐	☐
44 What is the common factor in all types of rainfall?	☐	☐	☐
45 Where and when is convectional rainfall likely to be most common in Britain?	☐	☐	☐
46 What type of weather does a tropical continental air mass bring to Britain in summer?	☐	☐	☐
47 Which air masses are most likely to bring snow and cold weather to Britain in winter?	☐	☐	☐
48 Where do hurricanes develop and why?	☐	☐	☐
49 Why do hurricanes weaken and their windspeeds fall as they pass over land?	☐	☐	☐
50 Why do anticyclonic weather conditions often lead to poorer air quality and sometimes to the formation of smog over cities?	☐	☐	☐
51 What is the greenhouse effect?	☐	☐	☐

Ecosystems

52 What is the difference between the biotic and the abiotic parts of an ecosystem?	☐	☐	☐
53 What important role do decomposers play in ecosystems?	☐	☐	☐
54 Why is nutrient cycling so rapid in tropical rainforests?	☐	☐	☐
55 Why are large areas of rainforest being cleared in some LEDCs?	☐	☐	☐
56 Why does deforestation result in a decline in the fertility of soils in tropical rainforests?	☐	☐	☐
57 How can deforestation affect the biodiversity of tropical rainforests?	☐	☐	☐

41 Weather describes the daily condition of the atmosphere in a place; Climate describes the average weather conditions over a period of time 42 Continental climate – large range of temperatures (cold winters, hot summers); Maritime climate – smaller range of temperatures (mild winters, warm summers) 43 Depressions – low-pressure weather systems bringing unsettled/changeable weather (cloud, wind and rain); Anticyclones – high-pressure weather systems in which air is sinking usually bringing settled, drier weather 44 Warm moist air rising 45 Inland areas in east and south-east England in summer, where temperatures are high leading to more warm, moist air rising 46 Warm/hot dry weather 47 Arctic and polar maritime air masses 48 Over the west Atlantic Ocean and Gulf of Mexico in late summer/early autumn where the oceans are warm (over 20 °C) – vast amounts of warm moist air rising to form the deep dense clouds of the hurricanes 49 When hurricanes pass over land they are deprived of their energy supply from the oceans 50 Sinking air in anticyclonic weather can prevent air pollution from dispersing 51 Greenhouse gases (carbon dioxide, methane and nitrous oxides) trap and reflect much of the wave energy radiating from the earth, keeping temperatures higher 52 Biotic – living parts of an ecosystem (plants and animals); Abiotic – non-living parts of an ecosystem (soil, water) 53 Decomposers break down waste or dead matter from plants and animals returning nutrients to an ecosystem via the soil 54 Dead matter decomposes rapidly in the hot and wet conditions 55 To use the land for logging, mining, cattle ranching and cash crop plantations 56 Deforestation removes the supply of nutrients through new humus and the heavy rainfall means that nutrients are rapidly leached from the soil 57

Population distribution and density

58 Why are some parts of the world densely populated?	☐	☐	☐
59 What factors might explain why some parts of the world have low densities of population?	☐	☐	☐
60 Why is the term 'population explosion' sometimes used to describe the changes in population shown in stage 2 of the Demographic Transition model?	☐	☐	☐
61 What factors cause birth rates to fall?	☐	☐	☐
62 Why do LEDCs and MEDCs have different shaped population pyramids?	☐	☐	☐
63 Why is the population of some MEDCs such as Germany and Sweden declining and what problems could result from this decline?	☐	☐	☐
64 How can 'push' and 'pull' factors be used to explain the causes of migration?	☐	☐	☐
65 What problems can rural-urban migration cause for cities in LEDCs?	☐	☐	☐

Settlement

66 What is the difference between the site and situation of a settlement?	☐	☐	☐
67 What is the difference between nucleated and dispersed settlements?	☐	☐	☐
68 In which parts of the world are cities growing most rapidly?	☐	☐	☐
69 What are 'squatter settlements'?	☐	☐	☐
70 How can 'self help' and 'site and service' schemes help to improve the quality of life in squatter settlements?	☐	☐	☐
71 What are the main features of inner city decline in MEDCs?	☐	☐	☐

58 High density – areas with good resources, transport, communications and industrial/economic development
59 Low density - areas with extreme climates (high or low temperatures, low rainfall), highland areas with steep slopes
60 Birth rates are high but death rates are declining due to improvements in healthcare and sanitation leading to a rapid rise in population 61 Lower infant mortality, use of birth control, rising prosperity, improvements in education, government policies 62 LEDCs have broader based pyramids due to higher birth rates, narrower at the top due to lower life expectancy; MEDCs have a narrower base due to lower birth rates and are broader at the top due to higher life expectancy 63 Due to low birth rates and small family sizes; Could lead to large elderly dependent populations needing to be supported by a smaller working population 64 Push – negative reasons persuading people to leave an area; Pull – positive reasons attracting people to move to a new area 65 Not enough jobs, housing and public services to support the rapidly growing population of these cities; Transport infrastructure unable to cope with the rapid growth 66 Site – the land on which a settlement is built; Situation – the location of a settlement in relation to its surrounding area (physical and human features) 67 Nucleated – clustered around a central point; Dispersed – scattered across an are 68 LEDCs 69 Unplanned settlements often built illegally and at very high densities by the very poor (usually migrants) in cities in LEDCs 70 They involve the squatter populations improving their housing conditions and provision of basic services by providing cheap materials and low interest loans; Site-and-service schemes provide building plots and services needed (roads, drainage, clean water, electricity and, in some cases, education and health services) 71 Loss of population and employment, environmental impacts of industrial decline (derelict land and buildings) and concentration of social problems associated with this decline

72 Why have urban regeneration schemes led to conflicts in some cities?	☐	☐ ☐
73 Why has slum clearance and the building of high-rise flats not always been successful in solving the problems of poor-quality housing in inner-city areas?	☐	☐ ☐
74 What is a 'green belt'?	☐	☐ ☐
75 Why are traffic problems increasing in many cities in MEDCs?	☐	☐ ☐
76 What is counter-urbanisation?	☐	☐ ☐
77 Why are out-of-town locations attractive for large shopping centres?	☐	☐ ☐

Classifying economic activities

78 What types of employment do you find in the following sectors of economic activity: a) Primary?　　b) Secondary?　　c) Tertiary?　　d) Quaternary?	☐	☐ ☐
79 What is the difference between subsistence and commercial farming?	☐	☐ ☐
80 What is the difference between intensive and extensive farming?	☐	☐ ☐
81 What is farm diversification?	☐	☐ ☐
82 What impacts does the removal of hedgerows have on the environment?	☐	☐ ☐
83 How has the Green Revolution helped to increase food production in some LEDCs?	☐	☐ ☐
84 What is a Trans-national Corporation?	☐	☐ ☐

72 Sometimes conflicts have resulted from schemes not benefiting local people (e.g jobs for commuters and housing too expensive for local people) 73 Clearing whole areas broke up communities; Difficult to generate community spirit in areas of high-rise flats; Low-cost design of flats needed to rehouse large numbers of people quickly – often fell into disrepair quickly; Placing high concentrations of families with social and economic problems in some flats made problems worse 74 Areas around cities (mainly countryside) protected from most types of housing, industrial and commercial development; Established to stop large cities from spreading and to preserve some open space for recreation 75 Increasing number of cars and dependence on cars; declining public transport services; old road networks that can't cope with the volume of traffic; concentration of employment and other activities in city centres leading to more people travelling into cities on a daily basis 76 People leaving towns and cities to live in smaller towns and villages in rural areas 77 Larger amounts of cheaper land upon which to build shopping centres; Often more accessible locations near ring roads and motorway junctions; Large suburban populations (more customers) 78 a) Extraction or collection of raw materials or food (farming, fishing, forestry and mining) b) Construction, manufacturing and processing c) Providing services d) Providing specialist information and expertise (e.g research, design, ICT, advertising and marketing) 79 Subsistence – producing enough food to feed a farmer and his/her family; Commercial – aims to make a profit from selling all or most of the farm produce 80 Intensive – large inputs of labour and capital per hectare producing high yields Extensive – often larger farms with lower inputs per hectare 81 Developing other activities outside farming to provide income 82 Loss of wildlife habitats; can also lead to an increase in soil erosion (removal of wind breaks) 83 Introduction of new high-yield varieties of rice, wheat and maize and new farming techniques (irrigation, agro-chemicals and machinery) 84 A very large company with factories, offices and branch plants in several countries

85 What does it mean when an industry is described as being 'footloose'?	☐ ☐ ☐
86 What factors have an important influence on the location of high-technology industries?	☐ ☐ ☐
87 What are the 'multiplier effects' of new industries locating in an area?	☐ ☐ ☐
88 What is the difference between the formal and informal sectors of a country's economy?	☐ ☐ ☐

Tourism

89 Why has tourism become the fastest growing industry in the last fifty years?	☐ ☐ ☐
90 What are the disadvantages of jobs in tourism?	☐ ☐ ☐
91 What benefits can tourism bring for LEDCs?	☐ ☐ ☐
92 What problems can tourism bring for areas in LEDCs?	☐ ☐ ☐
93 What are 'honeypot sites'?	☐ ☐ ☐
94 What is sustainable tourism?	☐ ☐ ☐
95 What are the advantages of eco-tourism?	☐ ☐ ☐

85 It is not restricted to particular locations (e.g raw material and energy sources) and can select the best location (e.g near to markets or skilled labour) 86 Location of research and development facilities, accessibility (near to motorways or airports), availability of skilled labour; attractive sites 87 Benefits of new business for other businesses and services in an area – increased demand for other products and services leading to more employment and more income to spend 88 Formal economy – 'official jobs' where workers are registered to pay tax; Informal economy – 'unofficial jobs' where workers are not registered to pay tax (often low paid with long hours and no employment rights) 89 More people with higher disposable incomes and longer paid holidays; Improvements in transport 90 Jobs in tourism are often seasonal, low paid and have low job security 91 Valuable source of foreign earnings; profits to fund other developments (transport, education, health and welfare); increasing awareness and understanding of other cultures and places 92 Environmental impacts (construction of hotels, transport routes, demand for water, pollution from sewage, etc); cultural impacts of mass tourism; concentration of benefits in tourist areas 93 Places attracting large numbers of visitors particularly at weekends and during holidays; Problems due to congestion, lack of parking and environmental damage 94 Tourism making sustainable use of natural resources without long-term damage to the environment 95 More sustainable; smaller numbers cause less damage to the environment; fewer negative impacts on local culture and communities; local communities benefit more from income from tourism; income can be used to improve health and education; encourages conservation of wildlife and the environment

Development

96	Why can energy consumption be used as an indicator of economic development?	☐ ☐ ☐
97	Why is a larger proportion of the working population of LEDCs employed in farming than in MEDCs?	☐ ☐ ☐
98	What is the 'balance of trade'?	☐ ☐ ☐
99	What problems can LEDCs face if they rely on the export of a narrow range of primary products?	☐ ☐ ☐
100	What are the three types of aid?	☐ ☐ ☐
101	What is 'tied aid' and how does it benefit MEDCs?	☐ ☐ ☐
102	What are the disadvantages of large 'top-down' development projects?	☐ ☐ ☐

Resources and energy

103	What is the difference between renewable and non-renewable sources of energy?	☐ ☐ ☐
104	What are the disadvantages of coal as a source of energy?	☐ ☐ ☐
105	What are the advantages of natural gas as a source of energy?	☐ ☐ ☐
106	What are the causes of acid rain?	☐ ☐ ☐
107	Why is international co-operation important if the problems caused by acid rain are to be solved?	☐ ☐ ☐
108	What are the disadvantages of nuclear energy?	☐ ☐ ☐
109	What are the limitations of wind power?	☐ ☐ ☐

96 As countries become more developed economically, the demand for energy increases 97 Farming is less mechanised in most LEDCs and more people rely on subsistence farming or the production of cash crops 98 The difference between the value of a country's imports and exports 99 It will be vulnerable to changes in demand and world prices for the products; If demand or prices fall, they will lose income and not have alternative sources of income to fall back on; If they rely on the export of agricultural products, they could suffer if there are bad harvests or environmental problems 100 Bilateral, multilateral and non-governmental aid 101 A donor country putting conditions on how the aid should be used, i.e that it is spent on products and services from the donor country 102 Often use expensive and complex technology increasing dependence on overseas suppliers; Not always appropriate for solving the problems of local people; Decisions often imposed from outside and don't involve local people 103 Renewable – can be used again to produce energy; Non-renewable – cannot be replaced once used up 104 High levels of air pollution (e.g CO_2) contributing to global warming Dangers involved in extraction (mining) 105 Cleaner (less pollution), more efficient (low waste) and easier to transport than other fuels 106 When fossil fuels are burnt, sulphur dioxide (SO_2) and nitrogen oxide (NO_2) are released as pollution; Sunlight converts these gases into sulphuric acid and nitric acid, which dissolve in moisture in the atmosphere, resulting in acidic rainfall 107 Pollution released when fossil fuels are burnt by industry in one country will be transported in the atmosphere to another country; These gases converted into weak acids dissolved in the atmosphere become acid rain falling on other countries 108 Concerns over safety, waste from nuclear fission is radioactive and needs to be reprocesssed or stored for a very long time. 109 Noise pollution and the visual impacts of large wind turbines; Variable strength of winds

Using maps in geography 1

Q Ordnance Survey (OS), sketch map, statistical map

Q 085218

Q 25 000cm

1 A choropleth map shows information using shading,
   e.g. population density. The darker the colour,
   the higher the value.

2 0720

3 075203

Using maps in geography 2

Q The highest land on the contour map
   is 200 metres above sea level.

1 Tatton Farm

2 a) public telephone
   b) church with a tower
   c) golf course

3 50 metres above sea level

Using data and statistics in geography

Q A graph, chart or table

Q 55 million

Q 150mm of rainfall in November

1 37 million

2 Your answer should include the following information:
   constant, high temperature all year round, little variation;
   highest temperature in the months of May, June, July
   and August;
   high rainfall during most the year, low rainfall during the
   hottest months (May, June, July August);
   highest rainfall in March (250mm); lowest rainfall in
   August (50mm).

Using images in geography

Q Clustered development of hotels along the coast, water
   sports and swimming pools

1 a) Possible land uses: tourist accommodation (hotel), water
      sports and recreation, transport
   b) Two possible sources of pollution: litter from tourists on
      the beach or motorboats and jet skis

2 a) Large cloud cover in north-west Europe, little or no cloud
      in southern Europe, e.g. Spain

Plate tectonics 1

Q The inner core, the outer core, the mantle and the crust

Q West

Q Two plates slide past each other

1 The Indo-Australian plate

2 Most of the world's major earthquake and volcano zones
   are found on or near plate boundaries.

3 The South American plate and African plate

Plate tectonics 2

Q Plates move apart, magma rises and cools, forms new crust.
   A build up of cooled magma may create underwater
   volcanoes and eventually volcanic islands.

Q Plates move towards each other, oceanic crust sinks and
   melts in the subduction zone to form magma.

Q San Andreas Fault, California, USA

1 A destructive plate boundary can be found where the Pacific
   and Eurasian plates meet.

2 Your answer should include the following information:
   Two plates meet and denser oceanic crust is forced
   underneath continental crust. Oceanic plate sinks and melts
   in the subduction zone to form magma. Magma rises up
   through cracks in the rock to form volcanoes. The friction
   created as the plates move may cause earthquakes.

3 Folding occurs when continental crust is pushed upwards to
   form fold mountains.

Earthquakes

Q Earthquakes occur when one of the earth's plates gets stuck,
   while moving. Pressure builds up, when this pressure is
   released an earthquake is caused.

Q Using the Richter Scale

Q Gas explosions and fires

Q People can prepare for earthquakes by: training people,
   buying earthquake kits, building earthquake-proof buildings
   and reducing the damage caused by an earthquake.

1 a) Focus: the point below the earth's surface where pressure
      is released causing an earthquake
   b) Epicentre: the point on the earth's surface directly above
      the focus

2 Your answer should explain that LEDCs tend to have less
   money and fewer resources to try and reduce the impacts of
   an earthquake, e.g. earthquake-proof buildings. They also have
   limited money to help rebuild an area and/or rehouse people
   after an earthquake. Also, with less money and fewer resources,
   it is difficult for the emergency services to cope with disasters
   such as earthquakes.

3 Your answer should include:
   The name of the example, the key facts about the
   earthquake (date and strength), key facts about the impact
   of the earthquake (lives lost, damage to buildings and roads,
   effects on the economy and the cost of repair and rebuilding).

Volcanoes

Q Magma that has reached the earth's surface becomes lava.

Q Magma chamber, main vent, secondary vent, crater

Q Blasts of hot air, gases and ash into the atmosphere, destructive flows of lava (lahars) that can destroy peoples homes, the landscape and claims lives

Q Advantages: rich fertile soil, health benefits and income from tourism

1 Volcanoes form when magma escapes from within the mantle. The magma forces its way through weaknesses in the earth's crust. As the magma rises, a build-up of pressure occurs and this causes a volcanic explosion. When the magma reaches the earth's surface, it becomes lava and cools to form new rock. After many explosions, this rock builds up to form a volcanic mountain. The process of folding (see p.18) can also form volcanoes.

2 Your answer should include: the name of a volcanic zone (e.g. areas of Japan, the Philippines or Iceland), an explanation of the different advantages that a volcano can bring people:
– Adding rich nutrients to the soil, which helps agriculture
– Income from tourism helps the economy
– Health benefits from hot volcanic springs
– Not wanting to leave the area or not having a choice about leaving.
You must explain why each point is important and develop your ideas.

Rocks and weathering

Q Igneous rock, sedimentary rock and metamorphic rock

Q Limestone is a permeable rock; clay is an impermeable rock

Q Acids in rainwater and river water dissolve the limestone. This creates distinctive landforms, many of them are found under ground, e.g. underground caves, stalagmites and stalactites.

1 Weathering involves the break up of rocks on the earth's surface. Erosion involves wearing away of the earth's surface.

2 Igneous rock is formed when magma reaches the earth's surface and cools to form new rock.

3 Weak acids in rainwater easily dissolve limestone. When a stream or river flows over limestone it dissolves joints in the rock and then disappears down into the rock and flows underground until it reaches an area of impermeable rock. This process of chemical weathering helps to form underground caves.

Glaciation

Q The UK, parts of the Alps, Scandinavia, New Zealand

Q A corrie is formed when snow and ice gathers in a hollow on the sheltered side of a mountain. The hollow becomes bigger because of freeze-thaw weathering. The hollow becomes a corrie.

1 Your answer should explain the processes of abrasion and plucking.

2 In your answer you should name a glaciated environment, e.g. the Lake District. You should then explain that glaciated landscapes are made up of very distinctive landforms and features. This makes them attractive to tourists and day visitors. Glaciated landscapes are also ideal for outdoor activities, e.g. hill walking, climbing and abseiling. However, many glaciated landscapes are national parks and are protected environments. This can cause conflict between those who want to visit and use the landscape and those who want to protect it.

The river system

Q One with inputs and outputs.

Q Surface runoff, throughflow and groundwater flow, channel flow.

Q The main river channel, watershed, tributaries, confluence, source and mouth.

1 a) A drainage basin is the area of land drained by a river and its tributaries.
b) A watershed is the boundary between two drainage basins, usually found on a highland region.

2 The movement of water through the soil.

3 Some water may fall directly into a river channel as precipitation, water may also reach the river by moving through the soil (throughflow) or by moving through the rock layer (groundwater flow). Some water may also travel overland towards a river (surface runoff).

Flooding and hydrographs

Q The amount of water passing through the river at a given point

Q Factors increasing the risk of flooding in a river: highland surrounding the river, lack of vegetation around the river and an urbanised landscape around the river

Q Lag time for graph A is 5 hours. Lag time for graph B is 10 hours.

1 River discharge may vary between two river basins for several reasons: the relief of the land around the river (influences runoff speed), the amount of vegetation surrounding the river (influences interception), the land use around the river (is it urban or rural)?

2 Peak discharge was reached after 16 hours.

3 Graph A shows river discharge for an urban area. Evidence for this includes the short lag time (5 hours) and the steep rising limb. This shows that water has reached the river channel quickly, which suggests that the area around the river is built up and urbanised, with little vegetation. It could also be surrounded by highland.

Flood management

Q Hard engineering tends to be more expensive and makes significant changes to the environment around the river.

Soft engineering options tend to be low-cost and have less of an impact on the natural environment.

Q Benefits: controlling the discharge and flow of the river, preventing flooding and generating hydroelectric power. Problems: dams can be expensive and they can reduce the amount of sediment in the river, which increases river erosion. Dams stop natural flooding and this can affect the quality of the soil around the river.

Q The planting of trees

Q Afforestation (planting trees), ecological flooding and land-use planning

1 Floods can destroy homes and claim lives. The cost of rebuilding and repairing the damage caused by a flood affects the economy of a country.

2 Your answer should include a discussion of hard engineering options (e.g. building a dam or modifying the river channel) and a discussion of soft engineering options (e.g. afforestation and changing planning regulations). Try and name rivers where some of these methods have been tried.

3 Your answer should include:
The name of the river studied and a description of its location.
Briefly explain the methods that have been used to help control flooding along the river, e.g. building a dam.
Explain the advantages of what has been done to the river for people, e.g. saving lives, generating hydroelectric power.
Explain the advantages of what has been done to the river for the environment, e.g. planting trees around the river.
Explain the disadvantages of what has been done to the river for people e.g. losing homes to build the dam.
Explain the disadvantages of what has been done to the river for the environment, e.g. preventing the natural flooding of the river.

River processes

Q The source of the river is where it begins.

Q Hydraulic action, abrasion, attrition and solution.

Q Traction is when large rocks and boulders are dragged or rolled along the riverbed. Saltation is when smaller pebbles are bounced along the riverbed. Suspension is when light material is carried along near the surface on the river. Solution is when minerals are dissolved in the water.

Q It may enter shallow water or an area of vegetation. The volume of the water may decrease.

1 Abrasion: small pebbles and rocks that are transported by the river wear away the riverbed.
Attrition: pebbles and rocks being transported by the river knock together and break down into smaller particles.

2 The size of the particles being carried by the river will decrease (get smaller) as the river moves from source to mouth. This is due to the process of attrition. Also, if the river loses energy it will deposit the largest rocks and pebbles first.

River features

Q V-shaped valleys, interlocking spurs, waterfalls and rapids.

Q In the lower course of a river, lateral erosion takes place. As the river erodes laterally, it swings from side to side, forming loops and bends. These bends are known as meanders.

Q In the upper course of the river, the channel in narrow, the gradient is steep and the main force of erosion is vertical. In the lower course of the river, the channel is wide, the gradient is gentle and the main force of erosion is lateral.

1 Diagram should include labels to show areas of erosion and deposition, the neck of the meander, the direction of flow and the line of the fastest flow within the river.

2 Your answer should include a named example, e.g. the Ganges Delta, and some brief locational information. Explain the advantages of living around the lower course of a river, e.g. flat land (good for farming, industry and construction), good access and potential for trade with other countries (ports) and rich fertile farmland.

River pollution

Q Domestic sewage, agricultural waste and industrial waste.

Q River pollution can lead to decreased oxygen levels and this means that many forms of river wildlife can not survive. Nitrate levels in the river may rise. This is called 'eutrophication' and can lead to more algae growth, which uses up oxygen and blocks out sunlight from the river.

Q It is difficult to manage the River Rhine because the river passes through many different countries, which makes it difficult to agree on who is responsible for polluting the river and who should pay to clear it up.

1 The impacts of water pollution in rivers include a loss of oxygen, which means less wildlife; an increase in nitrate levels, which means the growth of algae. River pollution can also be harmful to human health.

2 Industrial pollution from coalmines and steelworks, agricultural pollution (from fertilisers) and untreated sewage.

3 Your answer should include a named example of a river and some brief locational information. You should outline what has been done to reduce pollution and comment if possible on whether these methods have been successful. For example, to help control pollution along the River Rhine, ICPR was set up. This group was an international committee and they worked on different schemes to control pollution. Oxygen levels are now rising in the river and wildlife is returning to the river.

Water management

Q There is a difference between the regions of the UK that have good, reliable supplies of water and the regions that have the highest population densities. Most people live in the south-east of the UK while most of our water supplies are found in the north and west of the UK.

Q Building more reservoirs, transferring water by pipe to other areas, banning hosepipes during the summer and educating people not to waste water

1 Welsh

2 Your answer should include the names of places and/or countries where attempts have been made to manage water supplies and prevent water shortages. Your answer may include building new reservoirs, transferring water from one region to another through pipes, banning the use of hosepipes in summer and educating people not to waste water.

The sea at work

Q Swash is water washing up a beach.
Backwash is water washing back down a beach.

Q Deposition happens when waves lose energy and drop the material being carried

1 Waves are carried by the friction created when wind blows over the sea's surface.

2 A wave's energy is controlled by its strength, duration and fetch.

3 Fetch is the distance a wave travels.

4 Waves break into the beach at an angle due to the prevailing wind. Material is carried up the beach from X by swash. Material is moved back down the beach at right angles, by gravity, by backwash. Material is moved along the beach by each wave until it reaches Y. This process is called longshore drift.

Coastal landforms 1

Q Erosion is affected by the rate of erosion, rock type and structure, wave energy, human activity and amount of beach material.

Q A wave-cut platform is created when a cliff is attached by waves. Erosion creates a notch at the base of the cliff, which eventually cannot support the weight of the cliff above, which collapses, leaving a wave cut platform in front of the cliff.
A headland is attacked by waves on three sides. Weaknesses in the rock will erode first, making caves. These erode further into arches. The arches collapse with further attack to form a stack. This is then attacked by waves until only a stump remains.

Coastal landforms 2

Q Spurn Head, Humberside, is an example of a spit.

1 A 'notch' is associated with wave-cut platforms.

2 Hard rocks such as limestone are associated with headlands.

3 A = cave
   B = arch
   C = stack
   D = wave-cut platform

4 Formation of a spit:

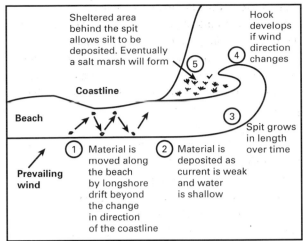

Sheltered area behind the spit allows silt to be deposited. Eventually a salt marsh will form

Hook develops if wind direction changes

Coastline

Beach

Prevailing wind

1 Material is moved along the beach by longshore drift beyond the change in direction of the coastline

2 Material is deposited as current is weak and water is shallow

Spit grows in length over time

Coasts and people

Q Coastlines are easily flooded and are fragile natural environments which can be easily destroyed.

Q Main coastal management strategies include groynes, sea walls, revetments, rock armour, gabions, offshore breakwater, beach nourishment and managed retreat.

1 Groynes are barriers placed at 90 degrees to the beach. Beach material collects against the barriers, creating a wider beach that slows erosion.

2 Beach nourishment is when material dredged from the sea bed is placed on beaches to create a barrier that slows erosion.

3 Hard management strategies are usually more expensive than soft management strategies. They are usually visually intrusive and less sustainable.

4 Some people believe that hard management techniques interfere with natural processes, causing more problems than they solve.

Mappleton case study

Q The Holderness coastline is on the east coast of Britain, south of Hornsea.

Q The Cliff Protection Scheme at Mappleton has had a knock-on effect further south along the coastline, speeding up the process of erosion from 2 to 10 metres per year.

Q Sue Earle is the farmer at Great Crowden, who had to move when the cliffs came within 4 metres of her home.

Q Your answer should include references to the news of local residents and businesses as well as local politicians and academics; some will be pro defences, others may be very against the protection scheme.

1 The main causes of erosion along the coast are strong waves, soft glacial till cliffs and longshore drift, which has removed the beach material protecting the cliffs.

2 Interventionist coastal protection schemes on one part of the coast may accelerate erosion elsewhere along the coast as the natural processes that take place to slow down erosion become imbalanced.

3 a) A coastline should be protected to save people's homes and businesses and transport links. A coastline should be left to erode naturally to maintain unique habitats such as spits, where the land behind the coast is of low economic value, and to maintain natural defences.

   b) Your answer should include views such as whether or not to protect a coastline is a dilemma. The needs of local people and governments, who are usually in favour of protection, need to be balanced against the needs of the whole coastline. For parts of the coast to remain protected naturally, erosion should be allowed to happen elsewhere as this will allow good beaches to build up in front of the cliffs and slow the erosion process. Other factors to consider include the cost benefit of protection.

Weather and climate

Q Weather is the actual condition of the air at a certain time and place. Climate is the average weather pattern for a place.

Q Tundra, mediterranean, maritime, tropical and hot desert are all world climate zones.

Q A continental climate has a large temperature range. Land is heated quickly by the sun and cools quickly as heat is lost to the sky at night.

1 The angle at which the sun's rays hit the earth determines temperature. It is colder at the poles than the equator because the sun's rays hit the poles at an oblique angle, diluting their intensity. At the equator, the sun's rays come from overhead, concentrating energy over a small area.

2 a) Berlin: 19 °C, Rome: 23 °C. They differ because Berlin has a central European (transition) climate and Rome has a Mediterranean climate.

   b) Berlin has a greater temperature range because it is a continental climate, whereas Shannon has a maritime climate.

   c) Shannon has the highest annual precipitation.

Weather and people 1

Q Three areas vulnerable to hurricanes are the Caribbean, South East Asia and the Indian subcontinent.

Q The hurricane passed through nine countries.

1 Hurricane Gilbert, the Caribbean

2 The course of Hurricane Gilbert:

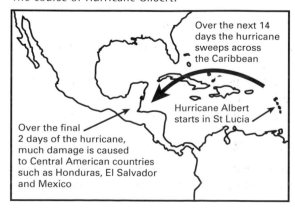

Over the next 14 days the hurricane sweeps across the Caribbean

Hurricane Albert starts in St Lucia

Over the final 2 days of the hurricane, much damage is caused to Central American countries such as Honduras, El Salvador and Mexico

3 Hurricanes happen in the Caribbean because it is a tropical region with warm sea. Hurricanes are most likely between May and November.

4 People can be affected in several ways; homes and businesses destroyed, no communications. People can be killed or injured by drowning or buildings collapsing, as well as being made homeless.

Weather and people 2

Q Semi-arid climates with long-term anticyclone weather systems experience droughts.

Q The Sahel region is south of the Sahara Desert in Africa; it includes the countries Mali and Sudan.

1 A drought is a long period of dry weather.

2 Humans contribute to drought by overusing existing water supplies and cutting down trees.

3 The Sahel is prone to drought because it is located in a semi-arid climate with continued anticyclone weather systems.

4 The nomads have been forced to move south into areas of savannah. These areas have now been stripped and the nomads have migrated to towns and cities, putting pressure on urban areas.

Climate change

Q The greenhouse effect:

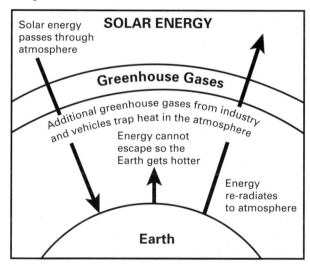

Solar energy passes through atmosphere

SOLAR ENERGY

Greenhouse Gases

Additional greenhouse gases from industry and vehicles trap heat in the atmosphere

Energy cannot escape so the Earth gets hotter

Energy re-radiates to atmosphere

Earth

Q The five main greenhouse gases are, methane, CFCs (chloroflourocarbons), nitrous gases, carbon dioxide and deforestation.

Q By reducing greenhouse gases – by planting trees and making sure countries cut emissions.

1 Sea levels are rising and an increase in global temperatures by 0.6 °C in 100 years.

2 Extra greenhouse gases created mainly by industry are accumulating in the upper atmosphere. They act like a blanket around the earth, trapping in heat that is being radiated from the earth's surface. This increases the temperature in the atmosphere surrounding the earth, causing global warming.

3   Three effects of global warming: rising sea levels, increasing global temperatures and changing weather patterns.

4   Britain will become wetter in winter but warmer in summer. Many low lying areas in southern England will flood permanently.

Understanding the weather

Q   Temperature is recorded in degrees Celsius on a thermometer. Precipitation is recorded in millimetres on a rain gauge. Air pressure is recorded in millibars with a barometer. Wind direction is measured with a wind vane and wind speed is recorded in km/hr on an anemometer. Relative humidity is recorded as a percentage using a wet and dry bulb thermometer. Clouds are recorded in oktas by observation.

Q   A prevailing wind is the most common wind direction.

Q   An isobar is a line joining places of equal air pressure.

1   A is fine, B is cloudy

2   A low pressure system or depression.

3   From the south-west

4   Thick cloud is associated with occluded fronts over the North Atlantic and the North Sea. These are caused by a large amount of warm moist air rising at these fronts, where a warm sector has been undercut by cold air moving in from a cold sector. A band of cloud is associated with the warm front over France. Warm air is rising over cool air. The cloud along the front is due to colder air undercutting warm air. This forces it to rise and cool, leading to condensation. Whirls of cloud associated with low pressure over the North Atlantic. Clear skies over Spain due to high pressure and no fronts over the Mediterranean.

UK weather

Q   Polar Maritime, Arctic Maritime, Polar Continental, Tropical Maritime, Tropical Continental

Q   Moist air rises, water vapour cools and condenses and forms clouds. Droplets grow in size and fall due to gravity.

Q   The UK is 50-58 °N of the Equator. The sun's rays are spread obliquely over the UK, giving cool temperatures. The North gets less sunlight and is cooler than the South. Places at higher altitudes have lower temperatures and more rainfall. The sea heats up and cools down slowly so places closer to the coast have less extreme climates with milder winters and cooler summers. Inland, the earth heats up and cools quickly, giving hotter summers but cooler winters. The Gulf Stream brings warm water (The North Atlantic Drift) from the Gulf of Mexico to the UK. The wind that blows over this water is warm, making winds from the west warmer. As this wind travels east across the UK, it cools, making the east coast of the UK colder.

1   Polar Maritime and Tropical Maritime

2   Tropical Maritime and Tropical Continental

3   The climate is wet due to relief rainfall and windy due to air masses travelling a long distance across the North Atlantic.

4   Much of the UK's rainfall is due to relief. Rain falls over the hills in the west. As air moves west to east it warms up. Therefore, the south and east of England are in the rain shadow and receive little rain.

Depressions and anticyclones

Q   A warm air mass and a cold air mass meet. The warm air rises to form a warm front. The warm front brings rain and then warmer brighter weather. Cold air behind the warm air forms a cold front. This brings colder and windier weather with heavy rain. The cold front moves quicker than the warm front, causing the two fronts to merge forming an occluded front which brings continual rain.

Q   An area of high pressure forms when cold air sinks and warms. The falling air results in more air at the surface, creating high pressure. Winds are light and blow out from the centre of high pressure clockwise.

1   Low pressure

2   During a summer cyclone

3   Nimbo stratus, alto stratus and cirrus

4   Cumulo nimbus and cumulus

5   As the depression approaches, there is rain followed by warmer, brighter weather. As the depression passes over, it turns windier and the rain is heavy and continual.

Ecosystems

Q   An ecosystem is an ecological system; a system of plants and animals that live together in a particular environment.

Q   Leaching occurs when nutrients are washed down into the lower layers of the soil where they can't be reached by plants or trees.

1   Any three from the key on the map, e.g. tropical rainforest ecosystems, coniferous forest ecosystems and deciduous forest ecosystems.

2   In the desert there is very little fertile soil (mainly sand) and there is little or no rainfall. This lack of inputs into the system means that the desert cannot support a lot of plants or animals.

3   Name an ecosystem and explain how energy flows through the ecosystem in the form of a food chain. Name the species of plants and animals found in the food chain.

The rainforest ecosystem 1

Q   Amazon Rainforest, Congo Rainforest and Indonesian Rainforest

Q   Inputs: sunlight, rainfall, moisture and nutrients from decaying leaf litter

Q   Logging, mining, cattle ranching or construction

The rainforest ecosystem 2

Q   Plant and animal species are lost, the nutrient cycle is broken and the soil loses fertility, soil erosion may take place, oxygen is lost and there is an increase in carbon dioxide in the atmosphere.

Q Encouraging sustainable logging, designating areas of the forest as protected areas, educating people about the importance of protecting the forest and encouraging mining companies to replant trees once they have finished mining the area

1 Found mainly along the equator, between the tropics of Cancer and Capricorn

2 The reason for the rich diversity of plants and animals in the rainforest is the high number of inputs into the system, e.g. sunlight and rainfall.

3 Your example should include the name of the ecosystem and some brief locational information. Explain why humans have changed the ecosystem and what impacts this change have had on the ecosystem. For example, humans have changed the tropical rainforest ecosystem because it is an important resource (providing wood, minerals and land). People have cleared the trees (deforestation) to use these resources. This deforestation has had many impacts on the forest ecosystem (loss of animal and plant species, loss of oxygen, and a loss of nutrients in the soil).

Population distribution and density

Q Population density = total population 4 the total land area (km²)

Q Densely populated = Japan
Sparsely populated = Greenland

Q The distribution of the UK is uneven. The areas of dense population density are found mainly in the south and south-east of the UK, as well as near major urban areas such as Manchester. There are few people living in Scotland and Wales and parts of northern England. These areas are sparsely populated.

1 The pattern of world population distribution is uneven. Most of the densely population areas are found north of the equator (in the Northern hemisphere). The most densely populated areas are found in Europe, North America and Asia. Areas that are sparsely populated include parts of central Australia, Canada and Northern Russia.

2 The distribution of the UK is uneven. Scotland, Wales and parts of northern England are sparsely populated. These areas are highland areas and they have a harsh winter climate. These conditions make farming and building difficult. Highland can also make communications difficult. The south-east of England is densely populated. This area has a mild climate and large areas of flat land. These conditions are good for farming and industry.

Population change

Q Births, deaths and migration

Q The Demographic Transition Model shows population change over time. It gives information about birth rates, death rates and the total population of a country. The model is divided into stages and as a country passes through the stages, the total population of that country increases.

Q LEDC pyramid: low life expectancy, MEDC pyramid: high-life expectancy.
LEDC pyramid: high birth rate, MEDC pyramid: low-birth rate
LEDC high number of young dependants, MEDC low number of young dependants
MEDC aging population

1 a) Around 3 billion
b) It had doubled to reach 6 billion

2 The Demographic Transition Model shows population change over time. As a country passes through the four stages of the model, the total population increases. This is because over time, birth rates and death rates rise and fall as a country develops. Death rates fall first as the country becomes wealthier, but birth rates remain high because people continue to have large families. This difference between birth rates and death rates is what causes the population to rise (high natural increase). During the final stage of the model a country will have low birth rates and death rates but a high total population.

Population issues

Q Children may be needed to earn money for the family, there may be limited contraception and limited advice about family planning, may be traditional to have large families. It may not be acceptable to use contraception.

Q In the village of Derada, a health centre was built where children can be vaccinated against disease and women can receive advice about contraception and family planning. Women are encouraged to plan their families and have less children.

Q Better healthcare, better diet, a high standard of living and good quality of life. Vaccinations given against disease, people are educated about not smoking and encouraged to exercise.

1 Due to a high rate of natural increase (high birth rates and falling death rates).

2 Your answer should include:
A named country or place that has tried to control population growth; this place may have tried to reduce or increase the birth rate. Describe what the country has done to control population growth and explain how these ideas or strategies changed the birth rate. For example, Derada, a Tanzanian village reduced its birth rate. A health centre was built in the village so that children could be vaccinated against disease and women could receive advice about family planning. This reduced infant mortality in the village and women could choose to have less children.

3 – Increased demand for healthcare and housing for elderly people. Taxes may have to increase to fund more

healthcare and services. Stress placed on the economy. The economically active sector of the population have to work harder to support the aging population.

+ Elderly people have important knowledge and skills to share with society and they can provide important votes for political parties.

Migration

Q Internal migration involves people moving within their own country.
International migration involves people moving from one country to another.

Q Push factors: unemployment, disease, famine, floods, a poor harvest
Pull factors: jobs, education opportunities, good healthcare, housing

Q The Turks couldn't speak German and faced racism and prejudice. When the German economy went into recession, the Turks were the first people to lose their jobs.

Q A refugee is someone who is forced to leave their home.

1 Your answer should include: the name of the county where people have migrated from and the place that they have migrated to. Explain why people have migrated from one country to another by listing the push and pull factors involved. Explain the impacts of this migration on both the country they have left and the country they have moved to.

Settlement

Q Site: the actual piece of land upon which a settlement is built
Situation: the position of a settlement in relation to the human and physical features surrounding it

Q Nucleated, linear and dispersed

Q A settlement hierarchy involves ranking settlements in order of their size and number

1 It is at the mouth of a large river, which means that it has a good supply of water and is in a good position for access and trade. The land around the river is flat, which is good for building homes and the development of industry

2 It is a linear shaped settlement.

3 It is a small cluster of buildings with no services.

Urbanisation patterns

Q A growth in the proportion of people living in urban areas compared to rural areas.

Q A millionaire city has over one million people living in it.
A mega-city has more than 10 million people living in it.

Q Push factors: poor farming conditions, unemployment, low wages, population pressure
Pull factors: employment opportunities, higher wages, good schools, healthcare services

1 The number of people living in urban areas is growing faster than the number of people living in rural areas.

2 These are found mainly in LEDCs and in the Southern hemisphere.

3 In LEDCs people are being pushed from rural areas due to unprofitable farming, low wages and a lack of social and leisure amenities. People are being pulled to urban areas because of good employment opportunities, higher wages and better social and leisure amenities, e.g. schools, hospitals and entertainment facilities.

Urban land-use patterns

Q Commercial and business, industrial, residential, open land

Q Poor-quality homes are found on the edge of the city in an LEDC. Higher quality housing is found near the centre of the city in an LEDC or on a major route out of the city. In an MEDC, poor-quality housing is found near the centre of the city and the high-quality housing on the edge of the city.

1 An urban land use model shows the typical pattern of land use in a city.

2 As you move away from the CBD, the value of the land falls. This is called distance decay.

3 Most MEDC cities have grown outwards from the centre. In the centre, land values are high and so buildings are built skywards. In the centre of the city (the CBD) the land is used for business, trade and administration. Around the CBD is the former industrial zone of the city. In many cities this area is now in decline and is a zone undergoing change. Around the edge of the former industrial zone is the poorest quality housing built during industrialisation for the workers. However, in some parts of this zone, gentrification has taken place and the housing has been improved. As you move out of the city, the quality of the housing improves. The suburbs on the edge of the city have lower density housing and larger homes with gardens.

4 In LEDCs the pattern of land use is different to MEDCs. Although LEDCs also have a CBD in the centre of the city, the lowest quality housing is found on the edge of the city. These are squatter settlements. High-quality homes can be found further inside the city or along the main routes out of the city.

Problems in LEDC cities

Q Urban growth in LEDCs is taking place because many people are moving from rural areas to the city. Also the rate of natural increase is high in cities (see p.83).

Q A squatter settlement is an unplanned, often illegal area on poor-quality land on the edge of a city with few services.

Q By improving access to services and improving peoples homes

1 People live in high densities in squatter settlements, they tend to have few amenities, e.g. running water, the waste disposal system may be poor and this can lead to disease. The squatter settlements tend to be far from the centre of the city where the jobs can be found.

2 Quality of life in squatter settlements may be improved by working with the residents of the area. Charities and governments may be involved in helping individuals and families. Some cities have introduced site-and-service schemes (people buy a piece of land connected to the main services of the city and build their own home). Self-help schemes can also be introduced whereby people are encouraged to improve their homes and gain ownership of the land.

Problems in MEDC cities

Q Unemployment, loss of traditional industries, out migration, closure of services, vandalism and crime

Q Advantages: new businesses attracted to the area, jobs created, investment in the area, improved housing and new housing
Disadvantages: can cause conflict between old and new residents living in the area, the new homes may be too expensive for the traditional residents of the area, the services may be too expensive for the traditional residents of the area

Q Old, narrow roads, increase in car ownership, commuting and bottlenecks where large roads on the edge of the city feed into smaller roads inside the city

1 Your answer should include: information about the history of the inner-city area (built during the Industrial Revolution to house factory workers), information about the decline of industry in the inner city and the problems that this then caused for the areas, i.e. the spiral of decline (loss of jobs, houses falling into disrepair, out migration and the closure of services).

2 The inner city can be regenerated or redeveloped. Redevelopment involves improving the physical environment of the area (clearing derelict buildings and refurbishing houses and flats), whilst regeneration tries to improve the area's economy and quality of life for the people living in that area (by creating jobs and improving services).

3 Your answer should include: the name of the urban area/s and a brief description of the problem affecting that area/s. Describe some of the ways in which transport problems have been managed in that area/s, e.g. encouraging people to use public transport, making public transport systems more integrated, building new roads and introducing park-and-ride schemes.

The urban–rural fringe

Q The outward growth of the city, whereby the city begins to spread into the countryside.

Q Rural areas are quieter than urban areas and have larger amounts of open space and greenery. Houses are larger and cheaper (for their size). This means that residents can enjoy a better quality of life by living on the edge of the city but still commute to work in the city, using trains or the motorways.

Q Out-of-town shopping centres encourage people to use their cars and this damages the environment. Out-of-town shopping centres can also take trade away from shops and businesses in the CBD.

1 Green belts were introduced to prevent urban sprawl (the outward growth of a town or city into the countryside).

2 The movement of people from urban areas to rural areas

3 Your answer should include the following points: large amounts of space and therefore plenty of room for expansion and car-parking, good access via motorways.

Classifying economic activities

Q Primary = farmer, secondary = food processing, tertiary = doctor, quarternary = computer programmer

Q Yes, in 'most' LEDCs, the majority of the population are employed in primary industry. In most MEDCs, the majority of the population are employed in tertiary industry.

Q Finished products, waste products and profits.

1 a) Teacher – – education
   Car assembly – manufacturing
   Doctor – health
   Web design – information technology
   Train driver – transport
   Builder – construction
   Logger – forestry
   Scientist – research and development
   Jeweller – retail

b)

Primary	Secondary	Tertiary	Quaternary
logger	car assembly builder	doctor train driver jeweller teacher	web design scientist

2 X = 70% Primary   10% Secondary   20% Tertiary
  Y = 25% Primary   45% Secondary   30% Tertiary
  Z = 5% Primary   25% Secondary   70% Tertiary

a) X is an LEDC

b) Z suggests de-industrialisation

3

Inputs
Raw materials (e.g. aluminium); labour; energy; capital; machinery; buildings → **Processes** Vehicle assembly; administration → **Outputs** Finished cars; waste products; profits

Feedback Reinvested profits

Classifying farming

Q Outputs are what is produced for sale, any waste and profits.
Q Arable farming = market gardening in UK

Pastoral farming	= dairying in UK
Mixed farming	= forest garden in Amazon
Subsistence farming	= rice growing in Bangladesh
Commercial farming	= dairying in UK
Extensive farms	= cattle ranches in South America
Intensive farms	= market gardening in UK
Nomadic farms	= herding across Sahel
Sedentary farms	= dairying in UK

1 Flat land, regular high rainfall, high temperatures, seeds/plants, large numbers of people, small quantity of machinery
2 Grazing, breeding, sheering, rearing, dipping.
3 Milk and dairy products for sale, animals for slaughter, waste products, profit, stock for breeding.
4 Goat herding across the Alps
5 A large farm with low yields per hectare. Inputs of either labour or technology will be low

Distribution of farming

Q Relief determines the type of farming the farmer can do. Steep land rules out arable farming which can only take place on fairly flat land.
Q Climate and relief
1 A farmer must consider the relief, soils and climate. Relief determines the type of farming and climate, which crops can be grown. Soil is also important as good soils favour arable farming.
2 A farmer must consider how easily his products can reach a market and the nature of that market. Other factors are the people and technology needed to make the farm a success. The amount of aid from or restrictions imposed by the TU may also be an important factor in making the farmer decide what type of farming to practise.
3 An arable farm is large, situated on fairly flat land, preferably south-facing. Soils need to be well drained, deep and fertile. The climate should allow a growing season with 60+ days at 6 °C or more. Rainfall should not exceed 1000m per annum. Arable farms use little labour, but many machines and some technology. Buildings are needed to store the produce. Most farmers will take full benefit of EU subsidies and allowances.

Changes in MEDC farming

Q Intensification is the process by which farms have become bigger and more productive.
Q Diversification is using land for non-farming purposes, e.g. setting up tearooms, campsites, B and B, farm shop.
Q Eutrophication is when nitrates are washed into water. These cause some plants and algae to overgrow, using up available oxygen, suffocating other forms of aquatic life, such as fish.
1 Farms are bigger, fields are bigger, trees and hedgerows removed, wetlands drained
2 CAP aimed to supply food to EU members at reasonable prices, protect farmers from non-EU competition and give them a fair price for their produce.
3 By introducing quotas to reduce surplus milk production and by applying 'set aside' to ensure arable farmers leave 15% of their land unused.
4 Nitrates are bad if they reach water supplies. They cause eutrophication and Baby Blue Syndrome.

Changes in LEDC farming

Q Mono-culture is when a single crop is produced, e.g. bananas.
Q HYVs are new high-yield varieties of rice, maize and wheat. They are smaller, faster growing and more pest resistant than traditional varieties.
Q 'Appropriate technology' is when equipment and skills are applied in LEDCs to improve production in a way that is sustainable for both people and the environment.
1 Most farming is subsistence, but can be either intensive or extensive. Some is nomadic. There is some commercial farming mainly on ranches and plantations.
2 To meet the needs of a growing population.

Changing secondary industry in MEDCs

Q A supply of raw materials close by, labour, fuel supply for power. More recently, proximity to markets and transport costs have become important location factors.
Q Iron- and steel-making are heavy industries.
Q Globalisation is the process by which businesses and national economies become integrated into a single global economy.
1 Footloose industries are free to locate where they want to. They are not over-dependant on any single location factor.
2 It has declined due to the growth of MNCs and globalisation. As the UK's raw materials have depleted, other countries have developed competitive industries selling cheaper products. This is because in the UK raw materials and labour costs are high. Goods can now be produced more cheaply outside the UK.
3 MNCs are symbols of globalisation because 40% of the world's trade is carried out by the 350 biggest MNCs.

Modern industry in MEDCs

Q High-tech industry is a footloose industry that makes high value products, from pharmaceuticals to consumer electronics.

Q Another example of an MNC is Sony.

1 Low-cost site, low rent, cheap labour for assembly, Greenfield site.
Attractive for high-skilled labour needed for R+D component of the industry, with links to associated industries

2 By giving financial help to companies wanting to locate in the UK. Making attractive greenfield sites available at low rents and making sure taxes are low.

3 By rebuilding the local community and economy following de-industrialisation. By employing and retraining large numbers of local people, and encouraging agglomeration economies to develop.

Changing secondary industry in LEDCs

Q A job would usually be temporary and without a contract. Workers are self-employed and do not have a regular income.

Q An example of an early NIC is South Korea and a new NIC is Thailand.

Q MNC's locate in LEDCs because labour is skilled and cheap, as are raw materials, and there are large markets to exploit.

1 Employment in the formal sector involves jobs with permanent contracts. Work in the informal sector does not involve contracts and is insecure.

2 It can be hard for an LEDC to industrialise because its people may be unskilled and the infrastructure undeveloped, and there may be little capital to make investments.

3 NIC stands for newly industrialising country. It is an LEDC that has been industrialised since the 1950s.

4 MNCs invest capital in LEDCs, and employ and train large numbers of people. Linked industries are developed, which set off the multiplier effect. Spin-offs include increased living standards and skills transferred to the LEDC.

Tourism

Q Anyone who spends one night or more away from their own home.

Q Your answer should include that tourism can be classified by resources – natural and man-made – or by duration, which is the length of visit or the distance travelled, whether it is local or international. Other explanations could mention location or activity.

Q Tourism generates wealth for investment and development in the host country.

1 People have more disposable income and paid holidays and more time to spend on a holiday. In addition, expectations and attitudes have changed and people expect to go on holiday at least once a year.

2 Tourism can bring economic benefits such as foreign exchange into a country and jobs. It can enable local cultures and traditions to continue and protect fragile environmental sites.

Tourism issues in MEDCs

Q UK tourism has diversified. As well as seaside tourism, recreational and urban tourism activities have been developed.

Q Pembrokeshire, Northumberland.

Q A honeypot is a site that is the focus of tourist activity in an area. It sometimes has shops and usually services like toilets and car parks.

1 UK tourism was facing stiff competition from the Mediterranean and traditional seaside resorts were in decline.

2 The Dales get many visitors because they are located north of the highly populated cities of Leeds, Bradford and Sheffield and east of Manchester. They are easily accessed by road.

3 Locals cannot afford to buy houses within the Dales, as rich city dwellers buy second homes to use at the weekend and for holidays. Honeypots like Malham are destroyed by too many visitors and tourist facilities. Local quarries damage the environment and views for visitors.

Tourism issues in LEDCs

Q An answer could include Thailand, Mexico, Sri Lanka, Kenya, Maldives, India, Nepal, Malaysia, Bali, etc.

Q Leakage is when the profits earnt from tourism go back to the MEDCs who run the tourist companies, and not into the local LEDC economy.

Q Ecotourism uses tourist resources today without damaging the environment for future visitors and local people.

1 They are attractive because of hot year-round climates, beautiful scenery, different cultures and they're cheap.

2 The multiplier effect refers to the positive spin-offs resulting from a tourist industry taking off in an LEDC. These would include increased local trade, developed infrastructure, and additional industry locating in the LEDC.

3 Tourism that uses resources today without damaging the environment and people for the future.

4 Benefits include increased wealth, better infrastructure and new jobs outside traditional jobs in primary industry. Costs include debasement of local cultures and traditions, large-scale environmental damage to fragile ecosystems and many jobs are informal and temporary.

Contrasting tourism 1: Spain

Q Yes, because there are abundant primary resources. In addition, there are many historic and cultural places to visit, plenty of entertainment and good infrastructure.

Q Warm, dry climate (May to November), long sandy beaches, rugged mountains

Q Mijas is located north of Fuengirola, just off the N340 road approximately 30km from the main town of Malaga.

Q Since the 1990s, holidays to LEDCs such as Thailand and the Caribbean have become cheaper and more accessible to the ordinary family.

1 The effects are both positive and negative. Tourism has improved the infrastructure and employment levels. It has also saved local communities and traditions from dying out. Some natural landscapes have also been protected from environmental damage. However, tourism provides mainly temporary and low-paid employment. Crime has increased and traditional lifestyles have been eroded. There has been much environmental damage, with ugly high-rise buildings and traffic congestion in resorts, litter and sea pollution, as well as pressure on limited water supplies.

2 The Costa Del Sol is responding by banning any more high-rise development, making resort centres more attractive and encouraging people to explore surrounding historic towns and countryside.

Contrasting tourism 2: Kenya

Q Mombassa has many natural features, such as a hot climate, beaches, wildlife and reefs as well as diversity of cultures and a reasonable infrastructure. All these factors make it a good location for tourism.

Q Tourism employs half a million people. Living standards of local people, including the Masai Mara, have improved through trading with tourists and the tourist industry. Greater foreign exchange has allowed Kenya to improve its infrastructure.

Q Civial unrest and crime have been widely reported in the western media This fact, alongside over-use of resources and over-commercialisation, has meant tourists have sought alternative safari destinations.

1 Primary resources are beaches, reefs, wildlife, and a hot year-round climate. Secondary resources are diverse cultures and good airport and road links.

2 Tourism has eroded traditional ways of life and displaced nomadic communities. Crime has increased as some resent the wealth of westerners. Many are worried about the impact of western lifestyles on the young, some are tolerant of the 'sun, sangria and sex' image of MEDC tourists.

3 By limiting numbers at reef and safari sites and limiting or refusing to allow any development in other coastal and inland areas

Development

Q More economically developed country; Less economically developed country

Q GNP, life expectancy, access to safe water, literacy rates, calorie intake, women's rights and infant mortality

Q The gap between rich and poor countries

1 a) GNP stands for gross national product – this is the amount of wealth produced by a country divided by the total population.

b) the adult literacy rate is the number of adults that can read and write, usually given as a percentage.

2 Your answer should include the names of both LEDCs and MEDCs that you have studied. Discuss the quality of life in MEDCs compared to LEDCs. Comment on some of the indicators used to measure development, e.g. GNP, life expectancy, adult literacy, healthcare, sanitation and education opportunities. Try not to be vague in your answer and learn some key statistics about the countries you have studied in class.

Trade and aid

Q LEDCs tend to sell primary products to MEDCs and buy manufactured goods from them. Primary products are low-value products, while manufactured products are high value. This means that LEDCs spend more on importing goods than they do exporting goods.

Q Trade systems were set up in colonial times, which favoured MEDCs. These trade systems still exist today, even though colonisation has ended. This means that LEDCs are dependent on MEDCs to buy primary goods off of them and sell them manufactured goods.

Q Fair trade is when companies and workers in LEDCs are paid a fair price for the goods that they produce.

Q Bilateral aid, multi-lateral aid and non-governmental aid

1 LEDCs tend to spend more on importing manufactured goods than they make selling primary goods – this leads to a poor balance of trade.

2 LEDCs are often dependant on MEDCs for trade and aid due to their colonial past. Trade systems were set up many years ago when many LEDCs were colonies ruled by MEDCs. These trade systems benefited MEDCs. Even though colonisation ended during the last century, many LEDCs still rely on selling their primary products to MEDCs and buying manufactured goods from them.

Development projects

Q The UN

Q Borrowing money, debt, disruption to local people and local communities, damage to the environment

Q Small-scale development projects are low-cost projects that involve local people and make use of local skills.

1 Your answer should include: a definition of large-scale development (these are high-cost development projects that often involve a high amount of foreign investment and changes to the infrastructure of a country). You should then explain what is involved in a couple of specific development projects. You must name specific places and explain what was done to help that country to develop, e.g. building the Three Gorges Dam in China, building the Trans-Amazonian Highway.

2 Your answer should include: a definition of small-scale development (these are small-scale, low-cost projects that involve local people and local skills). Name a specific development project in a specific place, e.g. the use of intermediate technology in Nepal. Explain why small-scale development has advantages for the people and environment of this place, e.g. the use of fuel-efficient cooking stoves in Nepal uses local skills and helps reduce the use of firewood, which protects the environment.

Resources and energy

Q Renewable energy sources will not run out and can be used again and again.
There are only limited supplies of non-renewable energy sources. They can not replaced once they have been used up.

Q Fossil fuels are the fossilised remains of dead plants and animals.

Q The global demand for energy is increasing as more and more LEDCs begin to develop and as MEDCs continue to use energy to power their industries and homes.

1 Coal burned to produce energy in the form of electricity. Coal can also be burned to produce heat for homes or it can be used to power trains.

2 Your answer should look at both people and the environment. Structure your answer by dividing it into two sections. Make sure that you discuss both the advantages and disadvantages of coal mining:
People: + creates jobs and a strong mining community. As a mining town develops, investment and services are attracted to the area. − When the coal supplies run low, the mine closes. This creates unemployment and services will be forced to close.
The environment: − mining creates traffic and destroys parts of the natural landscape. It is very difficult to re-landscape an old mining area. Coal produces waste material which must be stored.

Non-renewable energy

Q + Fossil fuels such as coal are relatively cheap and easily available in many countries, many fossil fuels burn efficiently.
− Fossil fuels are non-renewable, burning fossil fuels creates pollution and harms the environment, extracting fossil fuels such as coal can harm the landscape.

Q Acid rain can kill fish and other water wildlife, acid rain can damage and even kill trees, acid rain can also affect farming and agriculture.

Q Advantages: nuclear power is cleaner and more efficient than fossil fuels. It doesn't release greenhouse gases or cause acid rain.
Disadvantages: There are serious safety concerns after the accident at Chernobyl.

1 Pollution is generated which harms the environment, fossil fuels are non-renewable.

2 Acid rain is an international problem as it can be carried large distances by the wind. This means that many countries may be affected by acid rain. The country affected by acid rain may not be the country, that caused the pollution.

Renewable energy

Q Renewable energy sources will not run out and can be used again and again. Renewable energy tends to be more environmentally friendly.

Q Renewable energy sources can have problems and limitations: they are often expensive to research and develop, and often do not generate large enough amounts of energy to replace non-renewable sources, such as coal.

Q + Very clean source of energy, no waste is produced, a relatively cheap source of energy
− Wind turbines need to be placed in areas of open land, these are usually areas of environmental importance; wind turbines can be noisy and some people think they are unattractive.

1 Renewable energy sources are now being developed by many countries as people are concerned about non-renewable resources running out in the future. People are also developing renewable resources to help reduce environmental damage.

2 a) Oil
b) Use of fossil fuels may be reduced, whilst use of renewable energy sources may increase.

3 Your answer should include: the name of a renewable energy source, e.g. wind power, solar power or geothermal energy. Structure your answer in two parts:
List the advantages of using this source of energy (comment on their impact on people and the environment).
List the disadvantages of using this source of energy (comment on their impact on people and the environment).
Use the table on p.129 to help you with this question.

Last-minute learner

Plate tectonics

- Continental drift and the distribution of earthquakes, volcanoes and fold mountains can be explained by theories of **plate tectonics**. The earth's crust is made up of a series of plates that move around on the semi-molten **magma** of the earth's **mantle**. Huge **convection currents** in the magma rise towards the earth's surface causing the plates to move.
- Some plates move towards each other, some move apart and others slide past each other in opposite directions. These movements cause great stresses to build up within rocks at the edge of the plates.
- **Earthquakes** are the tremors created by a sudden, violent movement of the earth's crust.
- The shock waves from an earthquake will be strongest nearer to its **epicentre**.
- The damage caused by an earthquake can be divided into **primary effects** (immediate impacts, such as roads and buildings collapsing) and **secondary effects** (e.g. damage to communications and fires caused by gas leaks), which can be more destructive.
- **Volcanoes** occur when magma from within the mantle is forced upwards to the surface.
- Volcanoes can be classified by their shape and type of lava (i.e. **cone** and **shield** volcanoes). They can also be classified as **active**, **dormant** or **extinct** depending on how often and when they last erupted.
- Fold mountains are formed when two plates collide forcing the surface rock up into mountains.

Rocks and weathering

- Rocks are classified by the way they were formed, their composition and the geological time period when they were formed.
- **Igneous rocks** (e.g. granite and basalt) were formed when magma from inside the Earth cooled and solidified.
- **Sedimentary rocks** (e.g. sandstone, chalk and limestone) formed from fragments of other rocks or the remains of living things being compressed into rocks.
- **Metamorphic rocks** are igneous or sedimentary rocks that have been changed through intense heat and pressure (e.g. limestone becomes marble).
- Rock type (**geology**) influences the shape of the landscape. Stronger rocks form highlands. The **permeability** of rocks determines whether the landscape is wet or dry at the surface.
- **Weathering** is the breakdown of rocks at the surface or underneath soil without any movement. **Erosion** involves movement and is the wearing away of land by water, ice or wind.
- **Physical weathering** usually results from changes in temperature or pressure. **Chemical weathering** occurs when chemicals dissolved in water attack and break down rock surfaces.

Glaciation

- Glaciers weather and erode the landscape by **freeze-thaw weathering**, **abrasion** and **plucking**.
- Glaciers change the shape of the landscape, widening, straightening and deepening valleys.
- When ice melts, rock material is deposited by a glacier or by meltwater streams. **Glacial meltwater** picks up, rounds and deposits material. **Moraine** is the unsorted material carried and then deposited by a glacier. **Terminal moraine** marks the furthest point reached by a glacier.

The river system

- The **hydrological cycle** is the movement of water from the oceans to the atmosphere to land and back to the oceans. Some water is stored as surface water (rivers and lakes), in groundwater, in soil moisture and after interception on leaves. Water is transferred between different stores by evaporation, transpiration, precipitation, stem flow, infiltration, surface runoff (or overland flow), percolation, throughflow and groundwater flow (below the water table).
- The amount of water flowing past a particular point in a river over a given period of time is called the **discharge**.
- A **flood hydrograph** shows the relationship between precipitation and the discharge of a river. The time difference between the highest rainfall and discharge is called the lag-time. There is a greater flood risk when the discharge of a river has a short lag-time and a steep rising limb.

- There are six main factors that affect the way in which a river responds to rainfall – the amount and type of rainfall, rock type, soil type, land use, drainage density and the steepness of valley slopes.
- Rivers erode in four ways – **abrasion** (or **corrasion**), **attrition**, **hydraulic action** and **corrosion**.
- Rivers transport material by **traction**, **saltation**, in **suspension** and in **solution**.
- Rivers **deposit** their load when they do not have enough energy to transport it. The ability of a river to transport its load depends on its discharge.
- Most of the landforms in the upper part of a river result from erosion. The gradient of the valley is steep and friction from the bed and banks of the channel reduces the river's energy. The river mainly erodes downwards cutting a steep **V-shaped valley**.
- Features resulting from erosion include **potholes**, **rapids**, **waterfalls** and **gorges**.
- In the lower part of a river, the gradient becomes gentler and there is more **lateral erosion**. It has a wider and deeper channel, as more water is added from **tributaries**.
- Large amounts of silt, mud and sand are deposited where a river slows down or becomes shallower. This **alluvium** is also deposited on the floodplain on either side of a river when it floods.
- Features resulting from river deposition include **flood plains, levees, meanders, ox-bow lakes** and **deltas**.

The sea at work

- The power of waves increases as the strength of the winds and the distance over which the waves have built up (the **fetch** of the wave) increases.
- When waves break, water runs up a beach forming the **swash** and back down the beach as the **backwash**.
- **Constructive** waves are lower waves with a strong swash and weaker backwash that build up beaches. **Destructive** waves have a stronger backwash and erode material from a beach.
- Waves erode coastlines by **abrasion** (or **corrasion**), **hydraulic action**, **attrition** and **corrosion**.
- The type of rock and its resistance to erosion influences the way in which coastal landforms develop. **Headlands** form in harder rocks that are more resistant to erosion. Weaker rocks are eroded to form bays.
- Material is transported along a coast by **longshore drift**.
- Where a coastline changes direction (e.g. at the mouth of a river), longshore drift will continue to move sand, shingle and pebbles, forming a **spit**. Mudflats and salt marshes can build up behind spits in areas that cannot be reached by waves.
- **Hard engineering techniques** for protecting coastlines include the use of **sea walls, gabions, rock armour** and **revetments**. Sea walls deflect waves. Gabions, rock armour and revetments aim to dissipate or absorb the energy of waves.
- **Soft engineering techniques** aim to work with natural processes to protect coastlines. These include the rebuilding of beaches (**beach nourishment**) using sand dredged from further offshore. **Managed retreat** allows natural processes to take their course until the coast finds its own balance. Landowners near eroding coasts are moved inland and compensated for the loss of land.

Weather and climate

- **Weather** describes the daily condition of the atmosphere in a place.
- **Climate** describes the average weather conditions over a period of time.
- The world can be divided into **climatic zones** according to variations in temperature and the amount of precipitation.
- The **temperature** of a place is influenced by its **latitude, distance from the sea, height above sea level, aspect** and where the **prevailing winds** come from. Inland areas have a **continental climate** with a large range of temperatures (hot summers, colder winters). Areas near to or surrounded by the sea have a **maritime climate** with cool summers and milder winters.
- Britain has a cool, temperate (without extremes of temperature), maritime climate.
- Clouds form when warm moist air rises, cools and condenses to form cloud droplets. When the water droplets grow large enough, gravity causes them to fall as rain. Relief, convection and fronts are the main cause of moist air rising.
- The weather and climate of Britain is influenced by **air masses** – very large volumes of air of similar temperature and humidity (moisture content). The weather brought by an air mass depends on its origin and whether it passes over land or sea.
- **Depressions** are **low-pressure weather systems** that bring cloud, rain and wind. They form over the Atlantic where warm tropical air meets cold polar air and are responsible for the changeable, unsettled weather experienced by Britain over large parts of the year.

- **Anticyclones** are areas of **high pressure** in which air is sinking. They usually bring dry and settled weather, because sinking air warms up and can hold more moisture. Fog, mist, poor visibility and poor air quality can result from anticyclonic weather conditions.
- **Tropical cyclones** and **hurricanes** are very **intense low-pressure systems** that form over warm oceans (over 27 °C) usually in late summer/early autumn. They are very violent storms, with extremely strong winds and heavy rainfall.

- **Climate change** is a natural process. Human activity may be increasing the rate at which temperatures are changing. If they continue to rise, the world's climatic zones will be affected.
- Some scientists suggest that this **global warming** may be responsible for the increasing number and strength of extreme weather and climatic events, e.g. floods, hurricanes and droughts.
- **Greenhouse gasses** (carbon dioxide, methane and nitrous oxides) trap and reflect much of the long wave energy radiating out from the earth, keeping temperatures higher.

Ecosystems

- An **ecosystem** is the relationship between living (**biotic**) and non-living (**abiotic**) things.
- The different parts of an ecosystem are linked together by a series of **energy** and **nutrient** flows.
- Energy moves through an ecosystem by one member eating another in a **food chain**. **Food webs** are more complex, being made up of a series of inter-linked chains. During each stage of a food chain, energy will be lost through processes such as **respiration**.
- There are a number of important cycles in ecosystems, including the **water cycle**, **carbon cycle** and **mineral nutrient cycle**. The mineral nutrient cycle moves important minerals (e.g. nitrogen, calcium, phosphorous and potassium) through an ecosystem. Plants take up nutrients from the soil and use them to grow. **Decomposers** break down waste or dead matter from plants and animals, returning nutrients to the ecosystem via the soil.
- A **biome** is a large-scale ecosystem.

The characteristics of two of the world's major biomes:

	Hot desert	Tropical rainforest
Distribution (Where are they found?)	15-30 north or south of the equator	Between 5 north and south of the equator
Climate	Hot temperatures (up to 50 °C or more). High diurnal range (difference between day-night temperatures). Occasional heavy rainfall.	High temperatures (25-30°c) and high rainfall (over 2000mm per year)
Vegetation	Very sparse (except after heavy rainfall). Cacti and thorn bushes	Huge range of trees and other vegetation species
Biomass (average)	700 g/m2 Low biodiversity	90,000 g/m2 High biodiversity
Soil type	Little soil due to the absence of organic matter	Latosol Shallow and acidic Red upper layers due to presence of iron oxides. Thin relatively infertile soil due to the rapid nutrient cycling in the hot, humid climate
Wildlife	Limited range of species – insects, snakes	Abundant wildlife - huge range of species

Population distribution and density

- **Population density** is the number of people per square kilometre. Areas with good resources, high levels of economic activity, good transport and communication links are densely populated. Highland areas and areas with extreme climates (high or low temperatures, low rainfall) usually have low population densities.
- Populations change as a result of births, deaths and migration. The world population is growing rapidly, with the most rapid growth taking place in the poorer LEDCs. Most MEDCs have low rates of population growth with low birth rates in some countries even leading to a slowly declining population.

- The **demographic transition model** shows how the population of many European countries has changed over time. A **population explosion** occurred in the second stage, with death rates declining as a result of improvements in healthcare and sanitation. The rate of population growth declined towards the end of stage three as birth rates and family size fell as a result of family planning.
- **Population pyramids** show the numbers of males and females in different age groups in a population. Birth and death rates and migration influence the shape of a population pyramid.
- The **dependency ratio** compares the working population with the number of children (under 15) and elderly people (over 65) it supports as the dependent population.

LEDCs have large young dependent populations. The dependency ratios of MEDCs are growing as higher life expectancy leads to more elderly people needing to be supported with pensions, health and welfare services.
- In some countries, governments have introduced policies to control population growth, e.g. China's 'one child per family' policy. Improving education and family planning services has helped some LEDCs to reduce birth rates. In some MEDCs (e.g. Sweden) increases in family tax allowances and child benefit have been used to encourage people to have more children.
- **Migration** is the movement of people from one place to another.
- **Internal migration** – movement within a region or country.
- **International migration** – movement between countries.

- **Immigration** describes moving into a country, whereas **emigration** involves leaving a country.
- Push factors are the negative reasons forcing or persuading people to leave an area.
- Pull factors are the positive reasons attracting people to a new area.
- **Rural-urban migration** is common in many LEDCs and describes the movement of people from the countryside to live in towns and cities. In MEDCs there have been more movements of people away from cities to live in rural areas (**counter-urbanisation**).
- **Refugees** are people who have been forced to leave their homes as a result of wars, political or religious conflict, or due to environmental hazards, such as floods or famine.
- **Asylum seekers** are people who are asking to be recognised as a refugee and granted residency.

Settlement

- The **site** of a settlement is the land on which it is built. Site factors explain the reasons why a settlement was built in a particular place. The **situation** of a settlement describes its relationship with the surrounding area.
- Settlements can be classified into a **hierarchy** according to their size and the number of functions they provide.
- The **level of urbanisation** of a country or region is the proportion of its total population living in urban areas. MEDCs have high levels of urbanisation, but the fastest rates of urbanisation are occurring in LEDCs. The number of cities with a population of over one million in the world is increasing.
- The volume of rural–urban migration in LEDCs puts great pressure on housing, employment, transport and services (health, welfare, education, water, sanitation and power).
- **Squatter settlements** are the unplanned settlements often built illegally at very high densities by the very poor (usually migrants from rural areas) in cities in LEDCs. They are built from cheap materials, often on poor quality land, and lack basic services.
- **Self-help schemes** (e.g. providing low-cost materials, low-interest loans to set up businesses and improve healthcare and education) and **site-and-service schemes** are positive strategies that have helped to improve the quality of life for people living in squatter settlements.
- There are distinctive **patterns of land use** in cities including zones, sectors and smaller, irregular shaped areas or nuclei. In MEDCs the quality of housing tends to increase as you move out from the centre. In LEDCs higher quality housing is usually found just outside the CBD and poorer quality housing on the edge of the city.
- The **Central Business District** is the central area of a town or city where the main shopping areas, offices, financial services, public buildings and places of entertainment are located. It is the most accessible location because this is where the main transport routes meet. This means that large numbers of people can get into the CBD. As a result the demand for land is high, leading to high land values. These high land values lead to taller buildings to get more floorspace from the land available.
- **Inner-city areas** in MEDCs have been areas of decline, losing both employment and population. Young and highly skilled people have tended to leave these areas. **De-industrialisation** left large areas of derelict land. Over the last fifty years many different policies and schemes have tried to regenerate these inner-city areas (Comprehensive redevelopment, Urban Development Corporations, Enterprise Zones, City Challenge, Education Action Zones). From the 1980s, schemes have often involved partnerships between public and private investment. Much has been done to improve the environment and image of inner-city areas, but many problems remain.
- In an attempt to stop urban sprawl in the UK, **green belts** were set up around several cities in the 1940s. Most types of housing, industrial and commercial development were not allowed within the green belts. **New Towns** were built beyond the green belts to relieve congestion and overcrowding in the cities. Some were also set up in areas of industrial decline to provide new jobs and housing.

Classifying economic activities

- Economic activities can be classified according to the types of job people do – **primary** (extraction of raw materials and collection of food), **secondary** (manufacturing and construction), **tertiary** (services) and quaternary (specialist information and expertise, e.g. research and marketing).
- The employment structure of a country provides a good indication of its level of economic development; in LEDCs, a high proportion work in primary industries; in MEDCs, the highest proportion work in tertiary industries and there are more quaternary activities.
- Farms can be thought of as systems with inputs, farming processes and outputs. Farms can be classified in different ways – arable, pastoral and mixed; intensive or extensive; subsistence or commercial.
- Farming is influenced by a variety of physical factors (climate, soil and the slope or relief of landscapes) and human factors (distance from market, prices, farm subsidies, technology and farm ownership).
- Farming in the UK has experienced great changes In recent years. Technological, political and economic factors are now more important than physical factors.
- In parts of the UK, many arable farms have developed into large **agribusinesses** through the merger of farms and partnerships with big financial companies.
- Farming is still the main source of employment in many LEDCs. There are a wide variety of farming systems in LEDCs, from small subsistence farms and pastoral nomadism to large commercial plantations growing cash crops.
- Rapid population growth in many LEDCs has led to increased population pressure. Developments in farming through the **Green Revolution** (high-yield varieties of rice, wheat and maize, irrigation and new farm technology) have resulted in increases in food production.

- There are a number of important factors influencing the **location of manufacturing industries** – raw materials, power, markets, labour supply, transport, capital (investment), enterprise and government policy.
- **Heavy industries** (e.g. textiles, shipbuilding, iron and steel manufacture) in the UK developed during the Industrial Revolution of the late eighteenth and early nineteenth centuries. They located near to their raw materials because these were bulky and expensive to transport. Large factories needed large supplies of labour so industrial cities developed during this period.
- These heavy industries declined as the raw materials ran out or could be mined more cheaply from large opencast mines in other countries. **Deindustrialisation** resulted in high unemployment and environmental problems (derelict land and factories, polluted land, slag heaps)
- Successive governments have spent a lot of money trying to regenerate the economies of the older industrial areas of the UK, improving the environment and transport, and providing grants and tax allowances to attract new investment and employment.
- **High-technology industries** use advanced and often expensive techniques to design and manufacture high-value technologically sophisticated products, including consumer electronics, biotechnology, pharmaceuticals and medical equipment. They are **footloose** industries, usually involving the mass production and assembly of components. They often locate in clusters near to research and development facilities and sources of skilled labour (e.g. in science parks with links to universities).
- In recent years, a few very large **trans-national corporations (TNCs)** have taken control of industrial production and world trade. **Trans-national corporations** have offices, factories and branch plants in several countries. Their headquarters are usually in MEDCs or, in some cases, in Newly Industrialising Countries (NICs).

Tourism

- **Tourism** is one of the world's largest and fastest growing industries. This growth has resulted from improvements in transport, higher incomes (in MEDCs) and more people having longer paid holidays.
- The growth of tourism in many LEDCs has brought economic benefits through the **multiplier** effects of investment in tourist development (employment in construction, hotels, transport, tourist amenities and services), leading to improvements in infrastructure and public services.
- Mass tourism can have negative impacts on the environment and on local people and their culture.

- Ecotourism has been developed as an alternative to package holidays. **Ecotourism** is any form of tourism where the main attractions are the ecological resources (wildlife, environment and scenery) and human resources (traditional cultures, skills and buildings). It only caters for small numbers and the host community should be in control of and benefit from the tourism. It should be a more sustainable form of tourism without long-term environmental impacts.
- In the United Kingdom, twelve **National Parks** have been established to preserve and enhance the natural beauty, wildlife and cultural heritage of areas of outstanding natural beauty.

They aim to promote enjoyment of the countryside in these areas, as well as giving attention to the social and economic needs of local communities. There are **conflicts of interest** between different groups of people using the land in these National Parks.

- **'Honeypot sites'** are places of particular interest attracting large numbers of visitors. The visitor pressure at these sites can result in problems (e.g. traffic congestion and a lack of parking) and damage to the environment (e.g. footpath erosion).

Development

- The level of development of a country or region describes its wealth, level of economic growth and standards of living.
- An MEDC is a **More Economically Developed Country** (more wealthy). Most of the richer countries (MEDCs) are in the Northern hemisphere (apart from Australia and New Zealand).
- An LEDC is a **Less Economically Developed Country** (poorer or less wealthy). Most of the poorer countries (LEDCs) are in the Tropics and the Southern Hemisphere.
- Wealth (measured by **Gross National Product** per person) is not the only indicator of how developed a country might be. A variety of statistics can be used – life expectancy, infant mortality, birth and death rates, adult literacy, energy consumption, percentage of working population employed in farming, number of people per doctor. A variety of indicators need to be used when deciding how developed a country might be.
- **Colonialism** and continuing restrictions on trade help to explain why many LEDCs have stayed poor. Unstable government, a lack of investment and some natural factors (e.g. natural disasters) have also contributed.
- **International trade** involves buying goods and services from other countries (**imports**) and selling goods and

services to other countries (**exports**). The difference between the value of imports and exports is known as the **balance of trade**.
- Over 80% of world trade involves MEDCs. Many LEDCs rely on exporting a narrow range of primary products. Patterns of colonial trade created a state of **dependency** with LEDCs, relying on MEDCs to buy their primary products and to supply them with manufactured goods. This dependency has contributed to problems of **international debt**.
- **Aid** is any kind of help given to a country or a group of people to improve their quality of life (money, goods, equipment and technology, expertise and training). **Emergency relief** – short-term aid to solve immediate problems. **Long-term aid** aims to bring about long-lasting improvements to quality of life.
- There are three main types of aid – **bilateral aid** (one country to another), **multilateral aid** (from several countries through organisations like the UN) and **non-government aid** (including charities)
- **Top-down development** projects are usually expensive, large-scale projects with many aims (e.g. multi-purpose dam schemes). **'Bottom-up' projects** include small scale self-help schemes using local materials, **intermediate (appropriate) technology** and involving local people. They do not rely on expensive technology brought in from outside the area.

Resources and energy

- A **resource** is something that can be **made use of** by people. All industries rely on an input of resources. As a country develops and begins to industrialise, it uses more resources.
- **Natural resources** can be classified as **renewable** or **non-renewable**. Renewable resources will not run out: wind, sun, etc. Non-renewable resources are finite: coal, oil, natural gas, etc.
- Most of the world's **energy** is generated using **non-renewable fossil fuels**. Burning fossil fuels can have

a serious impact on the environment, including creating **acid rain** and contributing to **global warming**.
- **Nuclear power** is cleaner and more efficient, but there are serious safety concerns associated with it.
- Types of renewable energy include **wind power, solar power, tidal power** and **geothermal power**. All have their own advantages and disadvantages.
- The **Kyoto Agreement**, drawn up in 1997, is an attempt on the part of the world community to reduce emissions of carbon dioxide as a first step towards addressing the problem of global warming.